HANDBOOK OF
ANALYTICAL DESIGN FOR WEAR

HANDBOOK OF
ANALYTICAL DESIGN FOR WEAR

Editor: C. W. MacGregor

Authors: R. G. Bayer
T. C. Ku

Contributors: W. C. Clinton
R. A. Schumacher
J. L. Sirico
A. R. Wayson
C. W. Nelson

SPRINGER SCIENCE+BUSINESS
MEDIA NEW YORK
1964

ABSTRACT

This paper is the revised version of the "Handbook of Metal Wear Properties". In addition to presenting most of the information contained in that report, the current version also includes data about plastics, sintered metals, and platings, as well as a detailed procedure for wear design analysis. Section I briefly describes the engineering model for wear. Section II gives detailed descriptions and examples of design procedures for zero-and non-zero wear. Section III, through the use of figures and tables, presents quantitative information on material composition and the values of design parameters for various combinations of metals, plastics, and lubricants.

ISBN 978-1-4684-7169-4   ISBN 978-1-4684-7167-0 (eBook)
DOI 10.1007/978-1-4684-7167-0

Softcover reprint of the hardcover 1st edition 1964

The material contained in this Handbook was prepared by members of the staff of the Physical Technology Section, General Products Division, Development Laboratory, International Business Machines Corporation, Endicott, New York.

Library of Congress Catalog Card Number 64-8816

PREFACE

March 10, 1964

The problem of friction and wear between solid bodies is about as old as the human race. The early Egyptians and Romans had discovered the utility of lubricants in reducing friction and wear during a period many years B.C. From the fall of the Roman Empire until the Renaissance, little new information appeared. A major break-through occurred in establishing the laws of friction (friction independent of area and proportional to load) through the work of Leonardo da Vinci (1452 - 1519), Amontons (1699) and Coulomb (1785). While most of the studies until this time were based largely on a mechanistic approach, a new trend was initiated in the 1930's by F.P. Bowden and D. Tabor wherein the physics and chemistry of the problem were treated as well. Since then, a large literature has been built up dealing with such problems as metal transfer, molecular theories to explain wear, local welding between contacting surfaces, interlocking theories, wear-rate studies, the development of various test methods, effects of surface films, fretting phenomena, effects of temperature and environmental conditions, abrasion, surface energy relations, etc. The list of contributors in this field is indeed too long to cite here in any detail. Among those that might be mentioned, however, are R. Holm, G.A. Tomlinson, H. Ernst, M.E. Merchant, I.M. Feng, J.K. Lancaster, W. Hirst, F.T. Burwell, D.W. Baker, F.D. Brailey, E.R. Booser, E.H. Scott, D.F. Wilcock, E.E. Bisson, R.L. Johnson, M.A. Swikert, W.C. Clinton, K.H.R. Wright, H.M. Scott, K.L. Johnson, E. Rabinowicz, J.F. Archard.

The very complex nature of the wear process has in the past discouraged attempts to develop rational and numerical design procedures. While considerable success has attended such efforts in the creep of metals at elevated temperatures and to some degree in the fatigue of metals, the wear problem has not until recently been found to be too amenable to analytical methods. As a result, design against wear in machine parts has usually been given a rather nonchalant numerical treatment, and has been guided more by qualitative information on the reported propensities of different materials to resist wear rather than by quantitative considerations. Since the beginning of time, the total cost to the human race of parts which have failed due to wear must be staggering indeed. While wear has often in the past been the one most critical consideration in the design of many elements such as cams, gears, etc., recent trends in technology toward reducing maintenance, increasing the reliability, raising speeds and loads, and lowering the cost of products, make it even more important to have some way of treating the wear problem in a rational manner. The advent of computers has also rendered possible the detailed dynamic analysis of machines, thus supplying load data of use in such designs. Until now, then, we have not had available either a general wear model, applicable to widely differing conditions, nor the necessary detailed voluminous data for use in connection with such a model.

In order to fill this gap in our knowledge of the wear problem and to develop a practical engineering procedure to design rationally against wear, a long-range project was started in the Endicott Development Laboratory of IBM in 1952. The program has continued since that time. During the past three years, major attention has been given to compiling systematically a large amount of data covering many material and lubricant combinations subject to different kinds of wear. The experiments had to be conducted under well controlled conditions making use of highly specialized equipment. From an analysis of this data, it was possible to develop a wear model of wide application.

It should be emphasized again that this work has been developed for use by the design engineer. Sufficient information is available herein to help him solve many of his design problems whether for finite or negligible amounts of wear, lubricated or dry, whether the contacting parts are of bulk, layered or plated materials, and for solid or sintered metals, plastics, etc. It does not, therefore, purport to be a treatise on general wear phenomena, but is more of a practical design handbook. The material in the work has been used successfully by IBM engineers in designing over a hundred mechanical components since it was originally issued throughout the Corporation on April 10, 1962 under the title "Handbook of Metal Wear Properties". The material was also utilized in a course on Friction and Wear in the Advanced Mechanical Design School in Endicott, given by Advanced Technology personnel to a selected group of experienced design engineers. The present volume benefited from the exposure it has already had within the Corporation and was completely re-written adding new material on plastics, sintered metals, plated cases, etc.

Many persons have contributed to the project over various periods since its inception. It was started by J.E. Brophy, who was also responsible for its early administration. Support, advice and counsel were contributed by R. Walker, E. Garvey, E. Barber and J.J. Troy. Design contributions were added by L. Poch, R. Ingraham and others. R.B. Turner, S.C. Kingsley and J.Z. Devine gave assistance in obtaining part of the computational results and wear data. The authors were assisted in their preparation of this book by E.C. Deyo, who assisted in the preparation of the original drafts, and the members of the publications department of the Development Laboratory, who prepared the final manuscript.

The people most responsible for the success of the wear program and the contributors to the present work are all members of the staff of the Physical Technology Department of Advanced Technology in Endicott. They are R.G. Bayer, W.C. Clinton, T.C. Ku, C.W. Nelson, R.A. Schumacher, J.L. Sirico, and A.W. Wayson.

The project has been for several years under the direct supervision of T.C. Ku, who now also acts as a consultant on wear problems throughout the Corporation. It is with great pleasure that the writer now pays tribute to the contributions of all to this pioneering team effort.

C.W. MacGregor
Engineering Consultant and
Manager of Advanced Technology

General Products Division Development Laboratory
International Business Machines Corporation
Endicott, New York

TABLE OF CONTENTS

INTRODUCTION

The purpose of this work is to present to design engineers analytical design procedures for wear. For this purpose, the work has been divided into three main sections. The first section contains a short summary of the models used as a basis for the design procedures. The second section contains the design procedures and examples of their use. The third section contains a selection of tables and figures which are germane to the design procedures. This section contains a short summary of expressions for stress, values of Young's modulus and Poisson's ratio for the more common materials, and other tables, in addition to the main table of zero wear factors for a large number of combinations of materials and lubricants.

I ENGINEERING MODEL

The engineering model for wear developed by members of the Physical Technology Department has been discussed in several papers[1-4]. This section will not be a restatement all the concepts contained in those papers or of all the arguments and data which they present. It will be a summary of the basic conclusions reached in those papers, which are pertinent to design procedures.

A.   ENGINEERING MODEL FOR ZERO WEAR

Before the model itself is discussed, it is important that two terms be defined. The first term is zero wear. Zero wear is taken to be wear of such a magnitude that the surface finish in the wear track is not significantly different from the finish in the unworn portion. Roughly, this means that the depth of the wear scar is of the order of one-half the peak-to-peak value of the surface finish.

The second term is a pass. One pass is defined as a distance of sliding, W, equal to the dimension of the contact area taken in the direction of sliding.

The engineering model is concerned with the wear produced when two bodies are pressed together and relative sliding occurs. Basically the model states that wear can be controlled by limiting the maximum shear stress, τ_{max}, occurring in the vicinity of the contact region. More specifically, it states that wear can be held to a zero level for a particular number of passes, N, if τ_{max} is smaller than (or equal to) a certain fraction, γ, of the yield point in shear of the material, τ_y. In other words, if

$$\tau_{max} \leq \gamma \tau_y \qquad\qquad (I\text{-}1)$$

there will be zero wear for N   passes. The actual value of γ is dependent on the number of passes, N, the materials and lubricant used, and the lubricating condition. For 2000 passes, the values of γ are designated as γ_R. For quasi-hydrodynamic lubrication, γ_R ranges between a value of 1 and 0.54 depending on the degree of fluid lubrication occurring and the materials used.

For dry or boundary lubrication conditions, γ_R can take on one of two possible values, depending on the materials and lubricant used. They are

$$\gamma_R = 0.54 \qquad\qquad (I\text{-}2)$$

if the system has a low susceptibility to transfer, and

$$\gamma_R = 0.20 \qquad\qquad (I\text{-}3)$$

if it has a high susceptibility to transfer. The term system is defined as the combination of materials and lubricant used, if any.

The model further states that for a given system the value of γ for any other number of passes, $N \geq 2000$, can be related to γ_R by the following equation*

$$\gamma = \left(\frac{2 \times 10^3}{N}\right)^{1/9} \gamma_R \qquad (I-4)$$

For dry and boundary lubrication conditions, the values of γ_R are determined experimentally for each combination of materials and lubricant.

The above presents the pertinent features of the engineering model which are needed for designing for zero wear.

B. ENGINEERING MODEL FOR NON-ZERO WEAR

Before the model is discussed, the concept of wear, as used in these models, will be briefly considered. Wear is taken to be a change in the surface contour, such as a groove formed in a flat surface or a flat spot produced on a curved surface. Zero wear is a change in the surface contour of less than or of the same order as the surface finish. Non-zero wear would be a change in the contour which is greater than the surface finish. The basic measure of wear is taken to be the cross-sectional area, Q, of a scar taken in a plane perpendicular to the direction of motion (Figure I-1).

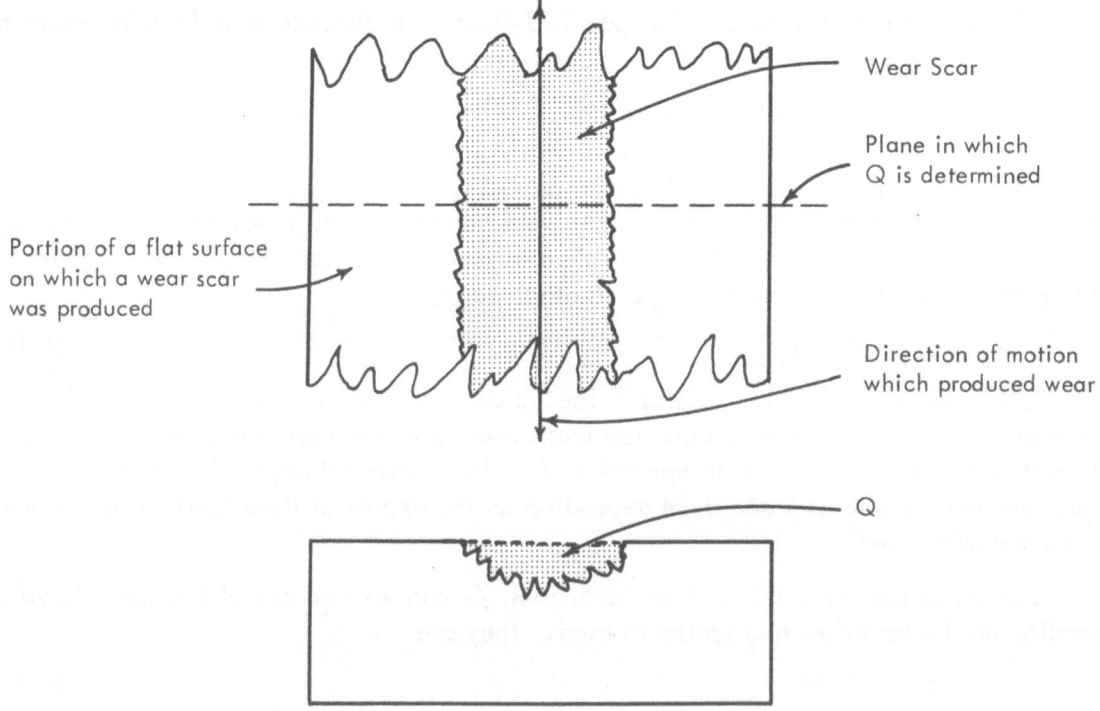

Wear Scar

Plane in which
Q is determined

Portion of a flat surface
on which a wear scar
was produced

Direction of motion
which produced wear

Q

Fig. I-1 – Diagram Illustrating Q, the Basic Measure of Wear

In the published papers concerning this model, this expression has been presented as an emperical one. However, more recent investigations have indicated that this equation can be derived from a rate equation describing wear, such as the one developed by Rozeanu, Wear, 6, (1963) 337 - 340.

The model for non-zero wear is formulated on the assumption that wear can be related to a certain portion, E, of the energy expended in sliding, and to the number of passes, N, by means of a differential equation of the type

$$dQ = \frac{\partial Q}{\partial E}\bigg|_N dE + \frac{\partial Q}{\partial N}\bigg|_E dN \qquad (1-5)$$

For design use, two equations developed from this representation are of significance. However, before they are given, it is necessary to distinguish between two types of wear, which will be designated as Type A and Type B. Type A wear is characterized by the predominance of transfer. This type of wear usually results in a quite rough wear scar, which gives evidence of materials having been plucked out of a surface. Type A wear usually occurs when metals, which have a tendency for transfer, are used under unlubricated conditions at shear stress levels which are close to or exceeding the yield points in shear of the metal. Type B wear is characterized by little or no transfer and with a relatively smooth wear scar.

The two equations derived from Eq. (1-5) are:

$$dQ = C' dN \qquad (1-6)$$

$$dQ = C'' (\tau_{max} W)^{9/2} dN + \frac{9}{2} \frac{Q d(\tau_{max} W)}{(\tau_{max} W)} \qquad (1-7)$$

Eq. (1-6) is appropriate for conditions under which Type A wear occurs and Eq. (1-7) is appropriate for conditions under which Type B wear occurs. The parameters, C', in Eq. (1-6), which is independent of N, and C'', in Eq. (1-7), which is independent of N and $(\tau_{max} W)$, may be dependent on such things as materials, lubrication, load, etc. which vary from case to case.

Integration of either of these two equations results in expressions which show how wear will progress with increasing numbers of operations of a mechanism. The actual manner in which such expressions are obtained and their use in design will be discussed in the next section.

II DESIGN PROCEDURES

The design procedures, which will now be described, are appropriate for cases of relative sliding motion between two bodies and are based on the models for wear discussed in Section I. These design procedures relate wear and design parameters in such a fashion that conditions for satisfactory wear behavior of a mechanisms can be analytically determined.

A. DESIGN PROCEDURE FOR ZERO WEAR

This procedure enables one to determine conditions which will insure zero wear for the desired lifetime of a given mechanism. The essential point of the design procedure is to insure that for each member of the mechanism, the following inequality holds

$$\tau_{max} \leq \left(\frac{2 \times 10^3}{N}\right)^{1/9} \gamma_R \tau_y \tag{II-1}$$

Satisfying this inequality insures that there will be zero wear for N passes. This expression can be used to determine whether a given set of design parameters will provide zero wear conditions for a given number of passes or to determine certain parameters in such a way as to insure zero wear. In either manner, the terms τ_{max}, N, γ_R and τ_y must be determined during the course of the design procedure. The ways in which these terms are obtained or determined are discussed below:

1.   Stress Calculation — The stress which must be calculated is the maximum shear stress occurring in the vicinity of the contact region, τ_{max}. This stress should not be computed simply on the basis of the normal load but should also include the effect of the friction force. For convenience, the formulations for several types of geometrical arrangements are discussed here. However, the applicability of the model is not restricted to the geometries discussed. If τ_{max} can be determined for any particular case, the model is applicable.

The first case discussed will be those geometries covered by the Hertz problem. The Hertz contact problem considers any two bodies which can be characterized by two principal radii of curvature in the contacting region, such as a sphere/plane, sphere/sphere, crossed cylinders, or parallel cylinders. It does not cover the case of conforming geometries, such as a sphere in a spherical seat of nominally the same radius, or a shaft running in a journal bearing which has the same nominal diameter as the shaft.

Parallel cylinders represent a limiting case in the Hertz problem and consequently there are special equations for this case. The expression for τ_{max} for the general Hertz case will be one of the following:

$$\tau \approx q_0 \left(\frac{1 - 2\nu}{3}\right) \frac{b}{a} \tag{II-2}$$

$$\tau \approx q_0 \sqrt{\frac{1}{4}\left[(1 - 2\nu)\frac{a}{a+b}\right]^2 + \mu^2} \tag{II-3}$$

$$\tau \approx q_0 \sqrt{\frac{1}{4}\left[(1 - 2\nu)\frac{b}{a+b}\right]^2 + \mu^2} \tag{II-4}$$

$$\tau \approx 0.31 \, q_o \tag{II-5}$$

where q_o is the maximum contact pressure, a and b are the semi-major and semi-minor axes of the ellipse of contact respectively, ν is Poisson's ratio and μ the coefficient of friction. Expressions for q_o, a and b will be given below. The expression appropriate for a given case is determined by the conditions of the problem. In general, for each problem one has to evaluate either Equations (II-2), (II-3), and (II-5), or (II-2), (II-4) and (II-5) and choose the highest value obtained in the group for τ_{max}. The first set of equations is used when sliding occurs in the direction of the semi-major axis, i.e., the "a" axis. The second set is used when the sliding direction is in the direction of the semi-minor axis, i.e., the "b" axis.

In a good many cases, the determination of the suitable expression for τ_{max} can be done in a shorter way. For combinations of materials for which $\nu > 0.035$ and $\mu < 0.30$, Equation (II-5) gives τ_{max}. If $\nu > 0.035$ and $\mu \geq 0.30$ either Equation (II-3) or (II-4) is the appropriate one. Equation (II-3) is used if the sliding is in the direction of the major axis; Equation (II-4) if it is in the direction of the minor axis.

The expressions for q_o, a and b are given as follows:

$$q_o = \frac{3}{2} \frac{P}{\pi a b} \tag{II-6}$$

where P is the normal load.

$$a = m \sqrt[3]{\frac{3\pi}{4} \frac{P(k_1 + k_2)}{B + A}} \tag{II-7}$$

$$b = n \sqrt[3]{\frac{3\pi}{4} \frac{P(k_1 + k_2)}{B + A}} \tag{II-8}$$

in which

$$k_1 = \frac{1 - \nu_1^2}{\pi E_1} \tag{II-9}$$

$$k_2 = \frac{1 - \nu_2^2}{\pi E_2} \tag{II-10}$$

$$B + A = \frac{1}{2} \left(\frac{1}{R_1} + \frac{1}{R_1'} + \frac{1}{R_2} + \frac{1}{R_2'} \right) \tag{II-11}$$

where the subscripts 1 and 2 refer to bodies 1 and 2. E is Young's modulus and R and R' are the principal radii of curvature. A radius is taken to be positive if its origin is in the body such as the radius of a sphere; it is taken to be negative if it is outside the body, such as the radius of a spherical seat. The numbers, m and n, are related to the geometries of the bodies and are given in terms of $\cos \theta$ in Table III-1 in Section III. $\cos \theta$ is given by

$$\cos \theta = \frac{B-A}{B+A} \tag{II-12}$$

where

$$B - A = \frac{1}{2} \left[\left(\frac{1}{R_1} - \frac{1}{R_1'} \right)^2 + \left(\frac{1}{R_2} - \frac{1}{R_2'} \right)^2 + 2 \left(\frac{1}{R_1} - \frac{1}{R_1'} \right) \left(\frac{1}{R_2} - \frac{1}{R_2'} \right) \cos 2\psi \right]^{1/2} \tag{II-13}$$

where ψ is the angle between the normal planes containing the curvatures $1/R_1$ and $1/R_2$ (Figure III-1).

In the case of parallel cylinders, the expression for τ_{max} will be one of the following:

$$\tau \approx K q_o \sqrt{\frac{1}{2}^2 + \mu^2} \tag{II-14}$$

$$\tau \approx K \left(\frac{1+\mu}{2} \right) q_o \tag{II-15}$$

where K is a stress concentration factor, μ, the coefficient of friction, and q_o the maximum Hertz contact pressure. Eq. (II-14) gives τ_{max} when sliding is in a direction parallel to the axis of the cylinders. Eq. (II-15) gives τ_{max} when the sliding direction is at right angles to the axis of the cylinders.

The stress concentration factor K* which appears in Eqs. (II-14) and (II-15) above is introduced to correct the Hertz formulae for the effect of corners at the ends of a finite cylinder. An effective magnitude for this factor and ways of determining it will be discussed at the end of this subsection.

The expression for q_o is:

$$q_o = \frac{2}{\pi} \frac{P'}{b} \tag{II-16}$$

where P' is the normal load per unit length and b is the width of the contact area.

* A further discussion of this factor in addition to some approximate analytical expressions for it are contained in a paper presented by C.A. Moyer and H.R. Neifert, entitled "A First Order Solution For The Stress Concentration Present at The End of a Roller Contact," at the 18th American Society of Lubrication Engineers Annual Meeting, May 1963, New York, New York.

b is given by

$$b = 2 \sqrt{\frac{P' R_1 R_2 (k_1 + k_2)}{R_1 + R_2}} \qquad (II-17)$$

where the k's and the subscripts have the same meaning as before and the R's are the radii of the two cylinders respectively. The sign convention for the radii is the same as that used for the radii in the general Hertz case.

The conforming geometries are the next type considered. In general, the expression for τ_{max} is

$$\tau_{max} \approx K q_o \sqrt{\left(\frac{1}{2}\right)^2 + \mu^2} \qquad (II-18)$$

where K is a stress concentration factor, q_o is the uniform pressure over the conforming surfaces and μ is the coefficient of friction. The stress concentration factor K is introduced to account for the presence of corners or edges. q_o is equal to the normal load, P, divided by the projected area of the contact. The projected area of the contact is obtained by projecting the surface of contact on a plane, perpendicular to the axis of the normal load, P. If, for example, the geometry is a flat against a flat,

$$q_o = \frac{P}{A} \qquad (II-19)$$

where P is the normal load and A is the area of the smaller flat. If the geometry is a shaft in a journal bearing, in which the diameter of the shaft has the same nominal diameter as the bearing,

$$q_o = \frac{P'}{d} \qquad (II-20)$$

where P' is the load per unit length and d is the diameter of the shaft.

The stress concentration factor, K, used in some of the formulae given above depends on the shape of the corners and edges occurring and can vary over several orders of magnitude. In the case of parallel cylinders it is possible to analytically determine the value of K in terms of the radii of the cylinders and the radii of the edges (see footnote on p. 6). However in the case of conforming geometries it is usually not possible to do this. In such cases, for the purpose of the design procedure, an effective value of K for a particular mechanism can be determined experimentally by means of a controlled wear test which utilizes the same geometries as those which occur in the actual mechanism, but not necessarily the same material. For the test, it is required that the materials used have equal hardness, not necessarily that of either of the members in the actual mechanism, and, in addition, the value of γ_R for the combination used in the test should be known. The test is used to determine a set of conditions, involving load, number of operations and coefficient of friction, at which non-zero wear first appears on either one of the members in the test. The set of conditions is then used in conjunction with Eq. (II-1), taken as an equality, to determine K. It has been found, experimentally, that typical effective values of K for well-rounded curves are of the order of 2 or 3. For sharp curves, the value can be as large as 1000.

If it is not possible to use one of these two approaches, one should require that the corners and edges are well-rounded and assume a value of 3.

It should be kept in mind that the equations given above are, strictly speaking, applicable only for homogeneous media, i.e., constant E and ν. However, it has been found that for the degree of accuracy required by the design procedures, these formulae may be applied to layered materials under the conditions that the E's and ν's of the various layers are not too different.

If this condition is not satisfied, the stress can no longer be computed on such a simple basis; a more complicated analysis is required. Such an analysis has been reported by D. Barovich, S.C. Kingsley and T.C. Ku[5]. This analysis indicates that in the case of Hertzian geometries if the thickness of the layer is greater than "2a", (general case), or "2b" (parallel cylinders). The effect of the substrate may be ignored in determining the stresses in the layer. It also indicates that if the thickness of the layer is much smaller than "2a" or "2b", the effect of the layer in determining the stress in the substrate may be ignored.

For convenience, the formulae given in this subsection are also contained in Figure III-1.

2. Calculation of the Number of Passes — Conversion of normally specified lifetimes into number of passes for each member of a mechanism can be done in a straightforward manner. Quite often lifetimes will be specified in the number of operations, L, required, such as the number of oscillations, revolutions or strokes. However, if the lifetime is not specified in that manner but in time, such as number of hours, years or months of operation of the unit, one should first convert such a specification into number of operations, L, which will occur during that time. The next step is to determine the number of passes, n, each member experiences in a unit-operation. The product, nL, gives the number of passes, N, for the entire lifetime.

It is convenient at this point to classify the two members of a mechanism as an "unloaded" member and a "loaded" member. If a member is designated as the "unloaded" member, this means that there are zones of this member which experience a loading and unloading in the course of a unit operation. If it is designated as the "loaded" member, this means that the region of this member which experiences the contact with the other member does not experience an unloading in the course of a unit operation. For example in the case of a ball sliding back and forth on a plane, the ball is the "loaded" member; the plane, the "unloaded" member. Another example is a shaft rotating in a journal bearing with a load, P, applied to the shaft. The shaft is the "unloaded" member; the journal bearing the "loaded" member.

There are two sets of expressions for n. The first set is when complete unloading of the original contact area occurs. In this case, n for the "unloaded" member can be determined in the following manner. Consider the motion which the two members experience in the course of a unit operation. In particular, consider an element of the "unloaded" member, which will eventually experience contact, but which is not in the contact region at the start or at the end of a unit operation and in addition is not at a point where the direction of travel is reversed. Then n, for the "unloaded" member, is the number of times this element experiences a loading and unloading in the course of a unit operation. For the " loaded" member,

$$n = \frac{S}{W} \qquad\qquad\qquad (II\text{-}21)$$

where S is the length of the sliding distance which occurs in a unit operation and W is the width of the contact area in the direction of sliding.

The second case is when the original area is not completely unloaded, which in most cases would imply that the motion is oscillatory. In such a case, the unit operation should be taken to be a single complete oscillation. Then

$$n = \frac{S}{W} \qquad\qquad\qquad (II\text{-}22)$$

for the "unloaded" member and

$$n = 2/\text{oscillation} \qquad\qquad\qquad (II\text{-}23)$$

for the "loaded" member.

As examples of determining n, consider the two previous examples in more detail. In the case of the ball sliding back and forth on a plane, assume the maximum length of travel in one direction to be 0.25 inches, i.e., the length of a stroke; the diameter of the contact region is found to be 0.012 inches. If a stroke is taken as the unit operation, n_b for the ball would be

$$n_b = \frac{S}{W} \qquad\qquad\qquad (II\text{-}24)$$

$$= \frac{.25}{.012}/\text{stroke} \qquad\qquad\qquad (II\text{-}25)$$

$$= 28/\text{stroke} \qquad\qquad\qquad (II\text{-}26)$$

since this is the "loaded" member. For the plane, which is the "unloaded" member,

$$n_p = 1/\text{stroke} \qquad\qquad\qquad (II\text{-}27)$$

which can easily be seen by looking at Figure II-1. This figure schematically indicates the initial and final contact positions in a stroke. The shaded area indicates the zone of the "unloaded" member which will be considered in determining n_p. It is seen that as the ball moves from the initial to the final position, this zone undergoes one loading and unloading. Consequently, $n_p = 1/\text{stroke}$.

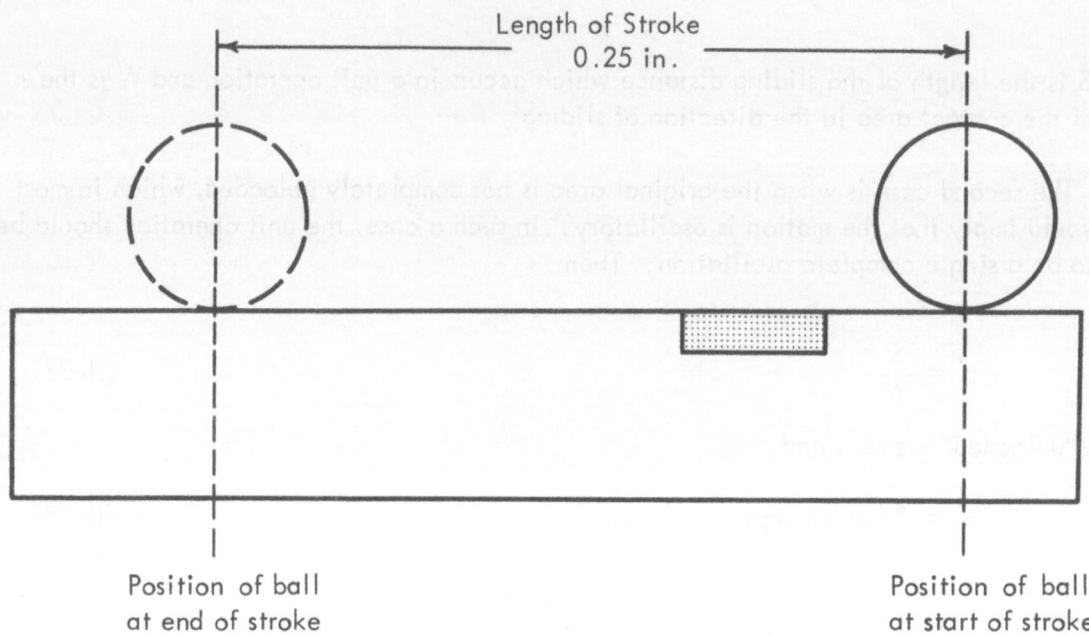

Fig. II-1 – Diagram Illustrating Determination Of n

In the second example, if the diameter, D, of the journal and bearing are nominally the same and if a complete revolution of the shaft is taken as a unit operation, the values of n are as follows:

For the shaft, which is the "unloaded" member,

$$n_s = 1/\text{revolution} \tag{II-28}$$

which can be seen by considering an element on the surface of the shaft which at the beginning of a rotation is outside of the contact zone. In the course of a single rotation, this element will first move into the contact zone, thereby being loaded once, out of the contact zone, thereby being unloaded once, and return to its original position at the end of the complete revolution. A single revolution thereby results in a single loading and unloading, and consequently $n_s = 1/\text{rev}$.

For the journal bearing, which is the "loaded" member,

$$n_B = \frac{S}{W} \tag{II-29}$$

$$n_B = \frac{\pi D}{\frac{1}{2}\pi D} \tag{II-30}$$

$$n_B = 2/\text{rev.} \tag{II-31}$$

assuming that the contact is made only over half the circumference.

3. <u>Determination of γ_R</u>   As was stated in Section I, γ_R is a function of the type of lubrication and the combination of materials and lubricant used. If the lubrication can be classified as quasi-hydrodynamic, γ_R can take on a maximum value of 1 or a minimum value of 0.54. The value which it will have in any given case is a function of the degree of hydrodynamic lubrication occurring, and also of the materials used in the mechanism considered. The actual value for a given mechanism should therefore be determined experimentally. However, if for a given mechanism the lubrication is known to fall in the quasi-hydrodynamic range, a safe value to assume for γ_R is 0.54.

If the lubrication can be classified as boundary or dry, i.e., no lubricant used, then γ_R varies with the combination. The value will either be 0.2 or 0.54, depending on whether the particular combination has a high or low susceptibility for transfer, respectively. The appropriate value for a given combination must be determined experimentally. For a large number of combinations, this has been done and the results are listed in Table III-3.

If a combination is not listed, γ_R may be estimated in one of two possible ways. The first way, and the safer way, is to assume a value of 0.2. The second is to find a combination which is listed and is similar to the one in question and then to use the value of γ_R for the listed combination. Two combinations are assumed to be similar if the lubricants and the materials of one combination are similar in composition to those of the other combination. The composition of the materials used for comparison should be that of the surfaces, and not necessarily that of the bulk metal, e.g., if the material has a case hardened layer on the surface, the composition considered should be that of the case hardened layer.

It is worthwhile to point out that if there is some question as to whether the lubrication is quasi-hydrodynamic or boundary, it is safer to assume boundary since γ_R for boundary conditions is always equal to or less than that for quasi-hydrodynamic conditions.

4. <u>Determination of τ_y</u> — If the yield point in shear, τ_y, is not known for a particular homogeneous material, it can be estimated from the microhardness of the material, H_m, which can be obtained by means of standard techniques[6]. τ_y is then determined from an experimental curve, given in Figure III-3, relating τ_y with microhardness, H_m. It has been found that such a curve is sufficient for the purpose of the design procedure. In the case of layered materials, such as platings and case hardened materials, τ_y varies with the distance from the surface inward and consequently no single value of τ_y exists. Additional considerations are therefore needed.

For an exact treatment of these cases, a knowledge of how the shear stress varies with the depth into the material as well as a knowledge of the behavior of the yield point in shear of the material with depth is needed. In such an approach, it would be required that Eq. (II-1) be satisfied at all points within the material. A simpler approach which is satisfactory for many engineering problems and which develops criteria which are on the safe side is now discussed.

For layered material where the layer on the surface is sufficiently thick, only this outer layer need be considered in the wear analysis. This layer is then considered in the same fashion as discussed before, i.e., as a homogenous material. In order for this approach to be satisfactory, the thickness of this layer should be as follows:

a)   0.005 inches or greater for conforming geometries;

b)   of the order of 2a or greater for geometries covered by the general Hertz equation;

c)   or of the order of 2b or greater for parallel cylinders.

The "2a" and the "2b" given above should be computed using the elastic constants of the component of the layered material which has the highest value of E.

In cases where the outer layer is not sufficiently thick, it is necessary to include both the outer layer and the substrate in the wear analysis. This is done by applying Eq. (II-1) to both the outer layer and the substrate. In such a case, the same values of N and τ_{max} are used for both layers. However, in the cases where the Hertzian geometries occur, τ_{max} is computed using the elastic constants of the layer which has the highest E.

The τ_y and γ_R used for the layers are, in general, different. γ_R for the outer layer is the γ_R appropriate for that layer, i.e., it can either be 0.54 or 0.2 depending on lubricant, etc. The γ_R for the substrate is taken as 0.54. The values of τ_y for both are determined from microhardness measurements. However, in this case a single value is not sufficient, as was true for homogeneous materials, but, rather, a hardness profile is required. This profile gives a measure of the variation of the hardness as a function of the penetration into the layered material. This profile is obtained by taking a series of microhardness measurements at various applied loads, each of these loads giving a different depth of penetration (Fig. II-2 (a)). This data is then used to plot a curve of hardness* vs. depth of penetration. A sufficient number of points should be used so that the hardness on the surface can be obtained by a means of extrapolation to zero depth of penetration. This usually requires that points be measured down to depths of penetration of the order of 20 μ inches. An example of such a profile is given in Figure II-2 (b). In such a curve, the transition from the surface layer to substrate is usually indicated by a change in the character of the slope of the curve, such as the change occurring around 20 to 30 μ inches, in the profile given in Figure II-2 (b).

---

*   The hardness, as obtained in this fashion, is more nearly the average hardness, $\bar{H}_m$, of the material up to that depth of penetration. However, it has been found that the hardness obtained in this manner is sufficiently accurate for the majority of engineering purposes. The hardness for a layer at depth "a" from surface can be related to H_m in the following

$$H_m (a) = \left. \frac{d\bar{H}_m(x)}{dx} \right|_a + \bar{H}_m(a)$$

where "a" is depth of penetration.

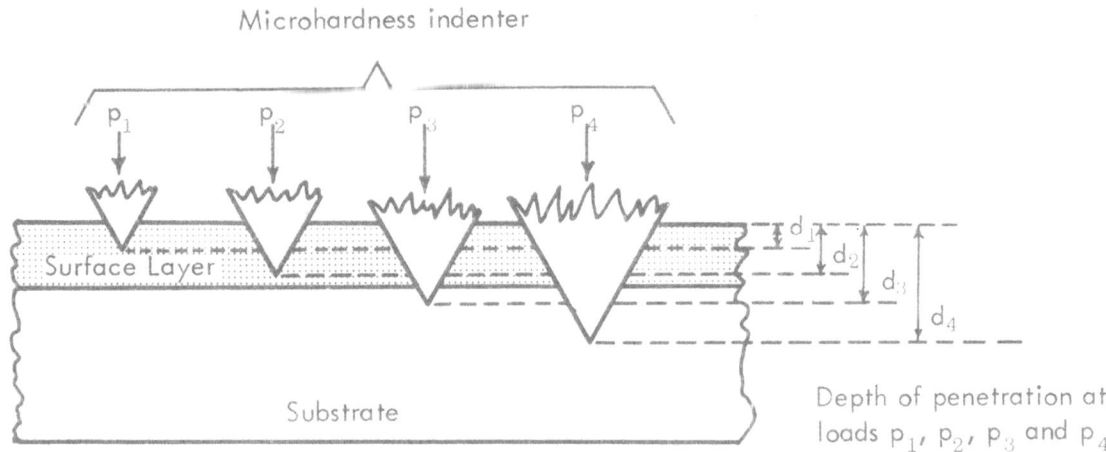

Fig. II-2 (a) - Schematic Illustrating Manner Of Obtaining Microhardness Profile

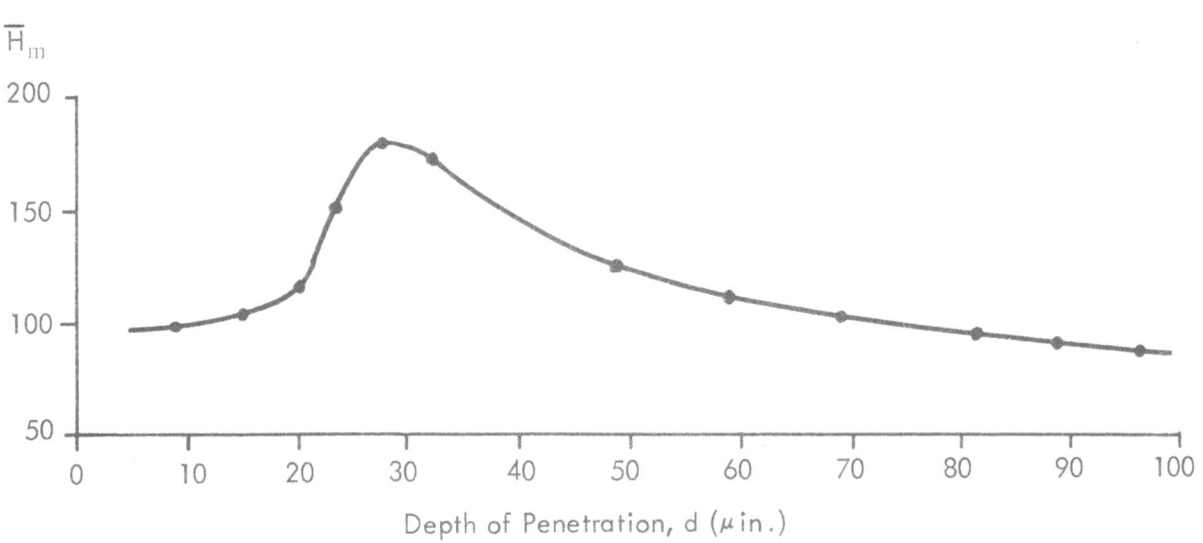

Fig. II-2 (b) - Average Microhardness Profile Of A Nickel-Cobalt Plating On Aluminum

After this change, the curve then usually monotonically approaches a region indicative of the hardness of the substrate, that is, a behavior similar to that seen from 30 μ inches on in Figure II-2(b). The τ_y taken for the surface layer is taken to correspond to the smallest value of H_m obtained before the curve monotonically approaches the substrate region. τ_y for the substrate is taken to correspond to the smallest value of H_m obtained in the region indicative of the substrate. τ_y is obtained from H_m in the same manner as before.

It should be kept in mind that the techniques for determining τ_y for and applying the wear criteria to layered media have been developed as a single procedure. As such, they have been shown to be satisfactory for use in the design procedure, even though neither is an exact treatment.

The above subsections have presented the ways of relating the quantities appearing in Eq. (II-1) to normal design quantities, such as load, geometry and materials. With these relationships and Eq. (II-1), it is possible to determine whether a given design will insure zero wear for the lifetime of a mechanism or not. Similarly, they enable one to determine a design for which there will be zero wear for a specified lifetime. The actual use of these relationships and Eq. (II-1) will now be demonstrated by a series of examples. The examples are grouped into two categories, the first being concerned with homogeneous material, the second being concerned with layered materials.

5. Examples

Homogeneous Materials

Example I

Consider the mechanism given in Figure II-3. The upper member is a fixed rod with a hemispherical end whose radius is 0.50 inches. The lower member is a cylinder of radius 2.00 inches, which is rotating about an axis parallel to, but displaced from, the axis of the cylinder. The load, P, is 1 oz., i.e. 0.0625 lbs. The hemisphere is to be made of 52100 steel with a through hardness of H_m 746; the cylinder is to be made of 8620 steel. The hardness of the 8620 is equal to H_m 216. Lubricant B (Table III-5) is to be used as a lubricant. This mechanism will be evaluated by means of the design procedure to determine whether or not there will be zero wear for 10^6 revolutions of the cylinder.

The criterion which must be satisfied for zero wear is

$$\tau_{max} \leq \left(\frac{2 \times 10^3}{N}\right)^{1/9} \gamma_R \tau_y \tag{II-32}$$

The first step in the evaluation of this expression is the determination of γ_R and the τ_y's. The type of lubrication is taken to be boundary and not quasi-hydrodynamic. Examination of Table III-5 shows that this particular combination of materials and lubricants is listed. The value of γ_R is seen to be 0.54. It is further seen that $\mu = 0.17$.

The τ_y's can be obtained from the graph relating hardness to τ_y (Figure III-3). In this case, the τ_y of 52100, H_m 746, is seen to be approximately 150 Kpsi; of 8620, H_m 216, τ_y is 40 Kpsi.

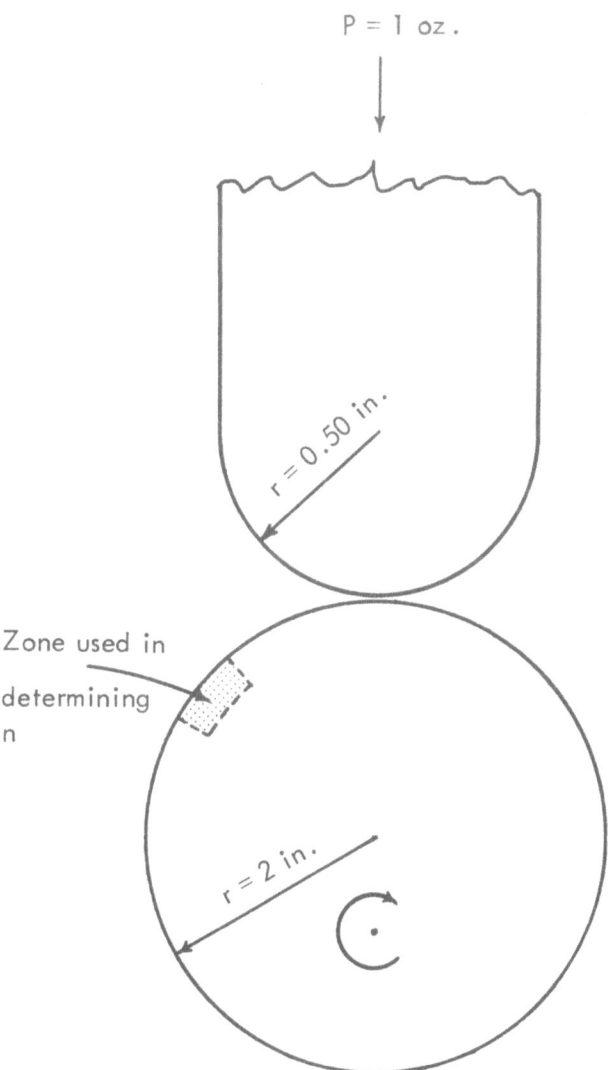

P = 1 oz.

r = 0.50 in.

Zone used in
determining
n

r = 2 in.

Fig. II-3 - Schematic of Mechanism Considered In Example I For Homogeneous Material

The next quantity which will be determined is τ_{max}. This particular contact is of the generalized Hertzian type with $R_1 = R_1' = 0.50$ inches, $R_2 = 2$ inches, $R_2' = \infty$ and $\psi = 0^0$. Since these metals are steels with a $\nu = 0.30$ and since $\mu < 0.31$, Eq. (II-5) gives τ_{max}, i.e.

$$\tau_{max} = 0.31 \, q_0 \tag{II-33}$$

The first quantities to be evaluated in calculating the stress are (B + A) and (B - A) where

$$B + A = \frac{1}{2}\left(\frac{1}{R_1} + \frac{1}{R_1'} + \frac{1}{R_2} + \frac{1}{R_2'}\right) \tag{II-34}$$

$$B + A = \frac{1}{2}\left(\frac{1}{.5} + \frac{1}{.5} + \frac{1}{2} + \frac{1}{\infty}\right) \text{in}^{-1} \tag{II-35}$$

$$B + A = 2.25 \text{ in.}^{-1} \tag{II-36}$$

$$B - A = \frac{1}{2}\left[\left(\frac{1}{R_1} - \frac{1}{R_1'}\right)^2 + \left(\frac{1}{R_2} - \frac{1}{R_2'}\right)^2 + 2\left(\frac{1}{R_1} - \frac{1}{R_1'}\right)\left(\frac{1}{R_2} - \frac{1}{R_2'}\right)\cos 2\psi\right]^{1/2} \tag{II-37}$$

$$B - A = \frac{1}{2}\left[\left(\frac{1}{.5} - \frac{1}{.5}\right)^2 + \left(\frac{1}{2} - \frac{1}{\infty}\right)^2 + 2\left(\frac{1}{.5} - \frac{1}{.5}\right)\left(\frac{1}{2} - \frac{1}{\infty}\right)\cos 2(0)\right]^{1/2} \text{in}^{-1} \tag{II-38}$$

$$B - A = 0.25 \text{ in.}^{-1} \tag{II-39}$$

Now m and n must be determined. This is done through $\cos\theta$ where

$$\cos\theta = \frac{B - A}{B + A} \tag{II-40}$$

$$\cos\theta = \frac{0.25}{2.25} \tag{II-41}$$

$$\cos\theta = 0.111 \tag{II-42}$$

Now from Table III-1 which relates $\cos\theta$ with m and n,

$$m = 1.08 \qquad n = 0.930 \tag{II-43}$$

"a" and "b" can now be determined. In this case, since both members have the same k's, they are given by

$$a = m\sqrt[3]{\frac{3\pi k P}{2(B+A)}} \tag{II-44}$$

$$b = n\sqrt[3]{\frac{3\pi k P}{2(B+A)}} \tag{II-45}$$

where

$$k = \frac{1 - \nu^2}{\pi E}$$

In Table III-2, it is seen that ν for steel is 0.30 and E for steel 30×10^6 psi. Consequently,

$$k = \frac{1 - 0.30^2}{\pi (30 \times 10^6)} \quad (\text{psi})^{-1} \tag{II-46}$$

$$k = 0.966 \times 10^{-8} \ (\text{psi})^{-1} \tag{II-47}$$

$$a = 1.08 \ \sqrt[3]{\frac{3\pi (0.966 \times 10^{-8}) (0.0625)}{2 (2.25)}} \ \text{in.} \tag{II-48}$$

$$a = 0.116 \times 10^{-2} \ \text{in.} \tag{II-49}$$

$$b = 0.930 \ \sqrt[3]{\frac{3\pi (0.966 \times 10^{-8}) (0.0625)}{2 (2.25)}} \ \text{in.} \tag{II-50}$$

$$b = 0.100 \times 10^{-2} \ \text{in.} \tag{II-51}$$

Having a and b, q_o can now be computed,

$$q_o = \frac{3}{2} \frac{P}{\pi a b} \tag{II-52}$$

$$q_o = \frac{3(0.0625)}{2\pi (0.116 \times 10^{-2}) (0.100 \times 10^{-2})} \ \text{psi} \tag{II-53}$$

$$q_o = 2.57 \times 10^4 \ \text{psi} \tag{II-54}$$

Consequently,

$$\tau_{max} = 0.31 \ (2.57 \times 10^4) \ \text{psi} \tag{II-55}$$

$$\tau_{max} = 7.95 \times 10^3 \ \text{psi} \tag{II-56}$$

The next step is to determine the number of passes experienced by the sphere and the cylinder during 10^6 rotations of the cylinder. In this case, the cylinder is the "unloaded" member and the direction of sliding is in the direction of the "b" axis of the ellipse of contact. Further, the contact area is completely unloaded. If we take a single, complete rotation of the cylinder as being the unit operation, n for the cylinder can be determined by considering the zone of the cylinder indicated in Figure II-3. In a single rotation, it is seen that this zone is loaded and unloaded only once. Therefore

$$n_c = 1/rev. \tag{II-57}$$

For the sphere,

$$n_s = \frac{S}{W} \tag{II-58}$$

$$n_s = \frac{2\pi(2)}{2(0.100 \times 10^{-2})} \ /rev. \tag{II-59}$$

$$n_s = 6.28 \times 10^3/rev. \tag{II-60}$$

The number of passes on the sphere and on the cylinder respectively are given by

$$N_s = 6.28 \times 10^3 \ L \tag{II-61}$$

$$N_c = L \tag{II-62}$$

since

$$N = nL \tag{II-63}$$

where L is the number of revolutions of the cylinder. For $L = 10^6$ revolutions,

$$N_s = 6.28 \times 10^9 \tag{II-64}$$

$$N_c = 10^6 \tag{II-65}$$

The inequality given in Eq. (II-32) can now be evaluated. For the cylinder, the expression is

$$7.95 \times 10^3 \overset{?}{\leq} \left(\frac{2 \times 10^3}{10^6}\right)^{1/9} (0.54)(4.0 \times 10^4) \tag{II-66}$$

$$7.95 \times 10^3 \leq 1.09 \times 10^4 \tag{II-67}$$

Therefore as far as the cylinder is concerned the conditions are satisfactory for zero wear. Now for the sphere the expression is

$$7.95 \times 10^3 \overset{?}{\leq} \left(\frac{2 \times 10^3}{6.28 \times 10^9}\right)^{1/9} (0.54)(1.50 \times 10^5) \tag{II-68}$$

$$7.95 \times 10^3 \leq 1.552 \times 10^4 \tag{II-69}$$

Therefore, since the inequality is satisfied for both the cylinder and the sphere, the specifications for this mechanism are satisfactory for having a zero wear condition for 10^6 revolutions of the cylinder.

If the specifications for the cylinder were changed to case hardened 8620, H_m 746, τ_y would change as well as γ_R. In this case, γ_R should be taken to be 0.2 since case hardened 8620 is closer in composition to 8660 than it is to 8620, and there is no information given in the Table for 52100-8660 combination.

Example II

In this example, the design of a journal bearing in which the nominal radii of both the shaft and the bearing are the same (Figure II-4) will be considered. In this case, the load, P, the length, ℓ, and the diameter, D, of the bearing are fixed at 5 lbs., 5 inches and 0.25 inches, respectively. The problem is to determine a combination of materials which will have zero wear under such conditions for a lifetime of 10^9 revolutions of the shaft.

Again, the criterion which must be satisfied to insure zero wear is

$$\tau_{max} \leq \left(\frac{2 \times 10^3}{N}\right)^{1/9} \gamma_R \, \tau_y. \tag{II-70}$$

This is to be satisfied for both the shaft and the bearing.

For the case of a journal bearing with a conforming geometry, the expression for τ_{max} is

$$\tau_{max} = \frac{KP}{\ell D} \sqrt{\left(\frac{1}{2}\right)^2 + \mu^2} \tag{II-71}$$

(see Eqs. II-18 and II-20). Two of the factors in this expression, K and μ, are unspecified in the statement of the problem. One of them, μ, is a direct function of the combination of materials and lubricant and cannot be specified independently of them. The other, K, can be specified to a first order approximation independently of the materials by specifying the geometry of the edges of the bearing. As an approximation, the edges of the journal will be specified to be well rounded and K will then be taken to be 3.

Therefore,

$$\tau_{max} = \frac{3(5)}{0.5 \times 0.25} \sqrt{(1/2)^2 + \mu^2} \text{ psi} \tag{II-72}$$

$$\tau_{max} = 1.20 \times 10^2 \sqrt{0.25 + \mu^2} \text{ psi} \tag{II-73}$$

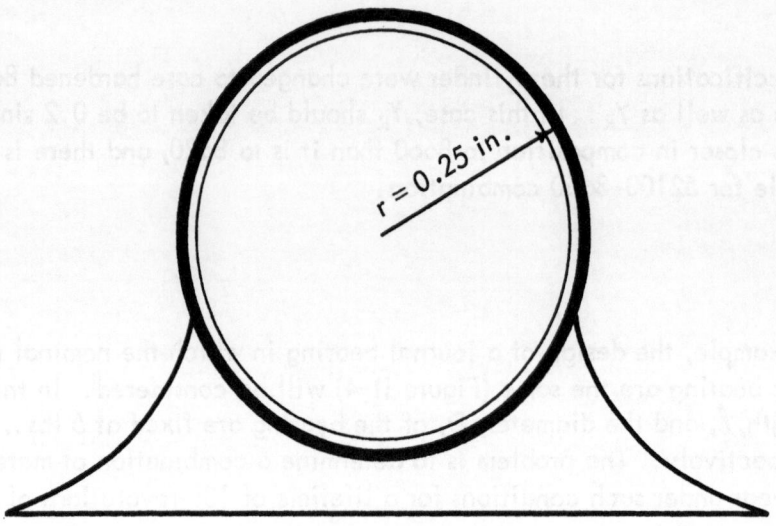

r = 0.25 in.

Fig. II-4 – Schematic Of Mechanism Considered In Example II For Homogeneous Material

Since in the case of a conforming geometry, N is solely a function of the geometry, N can be calculated from the information given. In this case, the shaft is the member which is unloaded fully, and if a unit operation is taken as a complete rotation of the shaft

$$n_s = 1/\text{rev}. \tag{II-74}$$

As was discussed earlier in the subsection on the determination of N, n_j for the journal is:

$$n_j = 2/\text{rev}. \tag{II-75}$$

As also was discussed earlier, the total number of passes for 10^9 revolutions is:

$$N_s = 10^9 \tag{II-76}$$

$$N_j = 2 \times 10^9 \tag{II-77}$$

since

$$N = n L \tag{II-78}$$

Consequently, any pair of materials and lubricant chosen must have a μ, Y_R and τ_y's such that

$$1.20 \times 10^2 \sqrt{0.25 + \mu^2} \leq \left(\frac{2 \times 10^3}{1 \times 10^9}\right)^{1/9} Y_R \tau_{y_s} \tag{II-79}$$

and that

$$1.20 \times 10^2 \sqrt{0.25 + \mu^2} \leq \left(\frac{2 \times 10^3}{2 \times 10^9}\right)^{1/9} Y_R \tau_{y_j} \tag{II-80}$$

or

$$\frac{Y_R \tau_{y_s}}{\sqrt{0.25 + \mu^2}} \geq 5.17 \times 10^2 \tag{II-81}$$

$$\frac{Y_R \tau_{y_j}}{\sqrt{0.25 + \mu^2}} \geq 5.56 \times 10^2 \tag{II-82}$$

where τ_{y_s} is τ_y of the shaft and τ_{y_j} is the τ_y of the journal.

The following combinations will be considered: delrin journal – 302 shaft ($H_m 270$), dry; 43 Aluminum ($H_m 51$) journal and shaft, dry; sintered bronze No. 1 journal – 52100 shaft ($H_m 746$), shaft, dry; and polystyrene journal – 302 shaft ($H_m 270$), dry. Of these combinations, the first, third and fourth are listed in Tables III-3, III-4 in the next section. The value of τ_y for the delrin is 1.235×10^3 psi; for the 302, 58×10^3 psi (from Figure III-3); for the 43 Aluminum, 8×10^3 psi (from Figure III-3); for sintered bronze No. 1, 22.5×10^3 psi; for the 52100, 150×10^3 psi; for the polystyrene, 0.535×10^3 psi. The values of μ and Y_R for delrin – 302, dry, are 0.25 and 0.54, respectively; sintered bronze 1 – 52100, dry, 0.27 and 0.2; polystyrene – 302, dry, 0.60 and 0.54. Since the second combination of 43 Aluminum against 43 Aluminum is not listed in the tables, Y_R and μ must be estimated. Since both members of the combination are the same, the susceptibility for transfer to occur would be expected to be high when there is no lubricant present. Consequently, Y_R should be taken as 0.2. When μ is not listed, it can be determined experimentally from other tables which list μ, or it can be estimated. In this case, μ is estimated to be about 1.0.

For these combinations, the left-hand expressions of Eqs. (II-81) and (II-82) are as follows:

Delrin – 302:

$$\frac{Y_R \tau_{y_s}}{\sqrt{0.25 + \mu^2}} = \frac{0.54 \times 58 \times 10^3}{\sqrt{0.25 + 0.25^2}} \text{ psi} \tag{II-83}$$

$$\frac{Y_R \tau_{y_s}}{\sqrt{0.25 + \mu^2}} = 56.1 \times 10^3 \text{ psi} \tag{II-84}$$

$$\frac{\gamma_R \tau_{yj}}{\sqrt{0.25 + \mu^2}} = \frac{0.54 \times 1.235 \times 10^3}{\sqrt{0.25 + 0.25^2}} \text{ psi} \qquad \text{(II-85)}$$

$$\frac{\gamma_R \tau_{yj}}{\sqrt{0.25 + \mu^2}} = 1.194 \times 10^3 \text{ psi} \qquad \text{(II-86)}$$

43 Aluminum – 43 Aluminum:

$$\frac{\gamma_R \tau_{ys}}{\sqrt{0.25 + \mu^2}} = \frac{0.2 \times 8 \times 10^3}{\sqrt{0.25 + 1.0^2}} \text{ psi} \qquad \text{(II-87)}$$

$$\frac{\gamma_R \tau_{ys}}{\sqrt{0.25 + \mu^2}} = 1.431 \times 10^3 \text{ psi} \qquad \text{(II-88)}$$

$$\frac{\gamma_R \tau_{yj}}{\sqrt{0.25 + \mu^2}} = \frac{0.2 \times 8 \times 10^3}{\sqrt{0.25 + 1.0^2}} \text{ psi} \qquad \text{(II-89)}$$

$$\frac{\gamma_R \tau_{yj}}{\sqrt{0.25 + \mu^2}} = 1.431 \times 10^3 \text{ psi} \qquad \text{(II-90)}$$

Sintered Bronze 1 – 52100:

$$\frac{\gamma_R \tau_{ys}}{\sqrt{0.25 + \mu^2}} = \frac{0.2 \times 1.50 \times 10^5}{\sqrt{0.25 + 0.27^2}} \text{ psi} \qquad \text{(II-91)}$$

$$\frac{\gamma_R \tau_{ys}}{\sqrt{0.25 + \mu^2}} = 52.8 \times 10^3 \text{ psi} \qquad \text{(II-92)}$$

$$\frac{\gamma_R \tau_{yj}}{\sqrt{0.25 + \mu^2}} = \frac{0.2 \times 2.25 \times 10^4}{\sqrt{0.25 + 0.27^2}} \text{ psi} \qquad \text{(II-93)}$$

$$\frac{\gamma_R \tau_{yj}}{\sqrt{0.25 + \mu^2}} = 7.95 \times 10^3 \text{ psi} \qquad \text{(II-94)}$$

Polystyrene - 302:

$$\frac{\gamma_R \ \tau_{y_s}}{\sqrt{0.25 + \mu^2}} = \frac{0.54 \times 58 \times 10^3}{\sqrt{0.25 + 0.60^2}} \ \text{psi} \tag{II-95}$$

$$\frac{\gamma_R \ \tau_{y_s}}{\sqrt{0.25 + \mu^2}} = 40.1 \times 10^3 \ \text{psi} \tag{II-96}$$

$$\frac{\gamma_R \ \tau_{y_i}}{\sqrt{0.25 + \mu^2}} = \frac{0.54 \times 0.535 \times 10^3}{\sqrt{0.25 + 0.60^2}} \ \text{psi} \tag{II-97}$$

$$\frac{\gamma_R \ \tau_{y_i}}{\sqrt{0.25 + \mu^2}} = 3.70 \times 10^2 \ \text{psi} \tag{II-98}$$

From these values, it is seen that the inequalities given in Eqs. (II-81) and (II-82) are satisfied by the first, second and third combinations, but not the fourth. Consequently, the first, second or third combination would be a satisfactory choice for the materials to be used in this mechanism.

Example III

In this example, the materials are fixed, but the geometry of one of the members of the mechanism is somewhat arbitrary. The mechanism is to be composed of a circular rod which slides back and forth across a supporting member (Figure II-5). The radius of the rod is 0.030 inches. The only requirement for the supporting member is that its axis be perpendicular to the axis of the rod and that its maximum width in a direction parallel to the rod be no greater than 0.100 inch. The two members are to be made of stainless steel; the rod of 440C, H_m 296, and the supporting member of 302, H_m 270. This mechanism oscillates back and forth and is to have zero wear for 10^{10} strokes under a normal load of 2/3 lb. The stroke length is 1/16 inch. It is further required that Lubricant A (Table III-5) be used as a lubricant.

Again, the expression which must be satisfied by both members of the mechanism is:

$$\tau_{max} \le \left(\frac{2 \times 10^3}{N}\right)^{1/9} \gamma_R \ \tau_y \tag{II-99}$$

For this particular combination of lubricant and materials, γ_R is found in Table III-3 to be 0.54, if boundary lubrication is assumed. Further, it is found from Figure III-3 that: 440C, H_m 296, has a $\tau_y = 63 \times 10^3$ psi; 302, H_m 270, has a $\tau_y = 58 \times 10^3$ psi. Consequently, Eq. (II-99) becomes for the rod and supporting member, respectively,

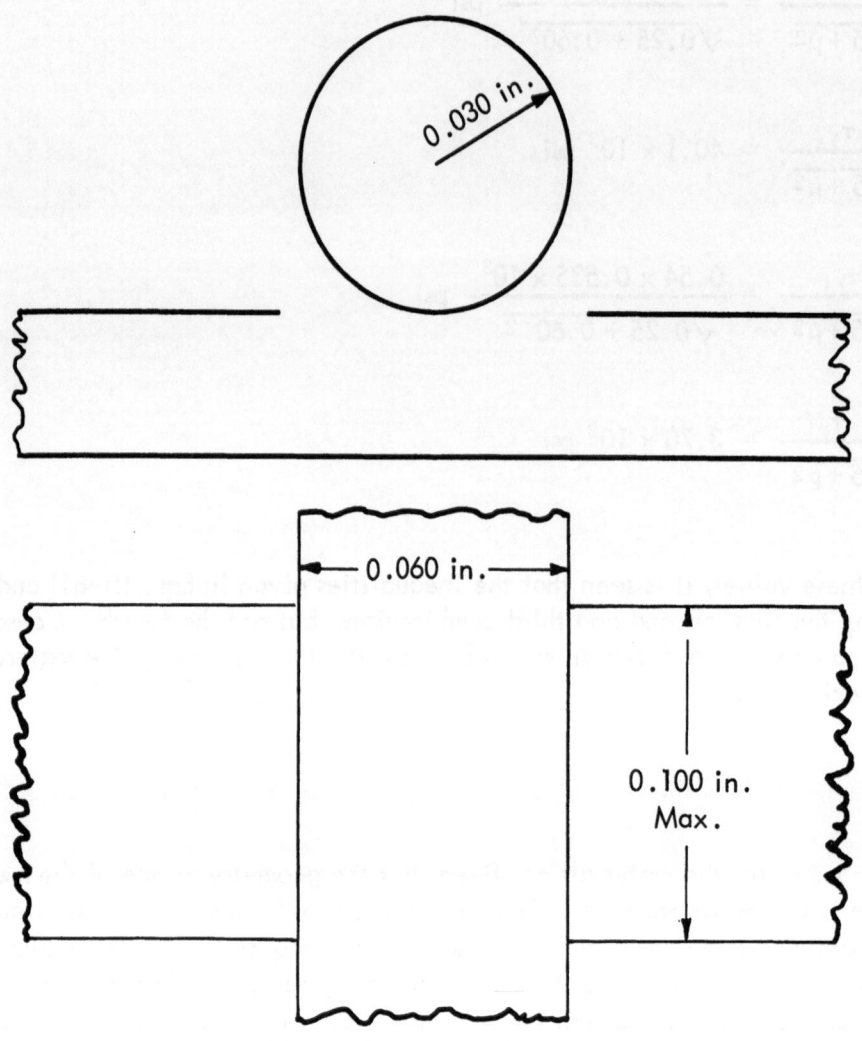

Fig. II-5 – Schematic Of Mechanism Considered In Example III For Homogeneous Material

$$\tau_{max} \leq \left(\frac{2 \times 10^3}{N_r}\right)^{1/9} (0.54)(63.0 \times 10^3) \text{ psi} \qquad \text{(II-100)}$$

$$\tau_{max} \leq \left(\frac{2 \times 10^3}{N_r}\right)^{1/9} (34.0 \times 10^3) \text{ psi} \qquad \text{(II-101)}$$

$$\tau_{max} \leq \left(\frac{2 \times 10^3}{N_s}\right)^{1/9} (0.54)(58.0 \times 10^3) \text{ psi} \qquad \text{(II-102)}$$

$$\tau_{max} \leq \left(\frac{2 \times 10^3}{N_s}\right)^{1/9} (31.4 \times 10^3) \text{ psi} \qquad \text{(II-103)}$$

The problem now reduces itself to solving Eqs.(II-101) and (II-103) for a geometry of the supporting member. Since the expressions for τ_{max} and the N's change for the different classes of geometries which might be used, it is impossible to do this on a straightforward basis. Intuition must be used. For example, in this particular case, it is easily seen that if the supporting member is a flat of the maximum width, τ_{max} would be lower than if the supporting member were a rod. Further, τ_{max} would still be lower if the flat had a groove in it which tended to conform to the radius of the rod. The equation used for τ_{max} in each of these cases is different. Consequently, the approach taken will be to first compute τ_{max} for the flat, assuming a stress concentration factor of 3, for well-rounded corners. If this geometry satisfies the Eqs. II-101 and II-103, the possibility of using another rod for the supporting member will be considered. If it does not, the possibility of using a flat with a groove in it will be considered.

The case of a cylinder on a flat is a limiting case of parallel cylinders, with one of the cylinders having an infinite radius. The equations appropriate for sliding in a direction parallel to the axes are used, which are

$$\tau_{max} = K\, q_0 \sqrt{\left(\frac{1}{2}\right)^2 + \mu^2} \qquad \text{(II-104)}$$

$$q_0 = \frac{2}{\pi} \frac{P'}{b} \qquad \text{(II-105)}$$

$$b = 2 \sqrt{\frac{P'\, R_1\, R_2\, (k_1 + K_2)}{R_1 + R_2}} \qquad \text{(III-106)}$$

In this case, $R_1 = 0.030$ inches, $R_2 = \infty$, $P' = \dfrac{0.667\ \text{lb.}}{0.10\ \text{in.}}$ and $k_1 = k_2$. The values of E and ν for stainless steel are obtained from Table III-2.

$$k_1 = k_2 = \frac{1 - \nu^2}{\pi E} \qquad \text{(II-107)}$$

$$k_1 = k_2 = \frac{1 - 0.30^2}{\pi\,(27 \times 10^6)}\ \text{psi}^{-1} = 1.073 \times 10^{-8}\ \text{psi}^{-1} \qquad \text{(II-108)}$$

$$b = 2 \sqrt{2 P'\, k\, R_1} \qquad \text{(II-109)}$$

$$b = 2 \sqrt{6.67 \times 2 \times 1.073 \times 10^{-8} \times 0.03\ \text{in.}} = 1.31 \times 10^{-4}\ \text{in.} \qquad \text{(II-110)}$$

$$q_0 = \frac{2}{\pi} \frac{6.67}{1.31 \times 10^{-4}}\ \text{psi} \qquad \text{(II-111)}$$

$$q_o = 3.24 \times 10^4 \text{ psi} \tag{II-112}$$

From Table III-3, μ is determined to be 0.15. Therefore,

$$\tau_{max} = 3 (3.24 \times 10^4) \sqrt{1/2^2 + 0.15^2} \text{ psi} \tag{II-113}$$

$$\tau_{max} = 5.07 \times 10^4 \text{ psi} \tag{II-114}$$

In this mechanism, the rod is the "unloaded" member and the support is the "loaded" member. Since the length of the stroke is 0.0625 inches and the size of the contact region in the direction of sliding is 0.100 inches, the rod is never completely unloaded. Consequently, the unit operation should be two strokes, or one oscillation. Therefore

$$n_r = \frac{S}{W} \tag{II-115}$$

$$n_r = \frac{2 \times 0.0625}{0.100} \text{ /osc.} \tag{II-116}$$

$$n_r = 1.250/\text{osc.} \tag{II-117}$$

$$n_s = 2/\text{osc.} \tag{II-118}$$

Therefore, since

$$N = n L \tag{II-119}$$

and since $\quad L = 0.5 \times 10^{10}$ oscillations

$$N_R = 0.625 \times 10^{10} \tag{II-120}$$

$$N_s = 10^{10} \tag{II-121}$$

Substituting the values for τ_{max}, N_R and N_s into Eqs. (II-101) and (II-103)

$$5.07 \times 10^4 \overset{?}{\leq} \left(\frac{2 \times 10^3}{0.625 \times 10^{10}} \right)^{1/9} (3.4 \times 10^4) \tag{II-122}$$

$$5.07 \times 10^4 \not\leq .645 \times 10^4 \tag{II-123}$$

$$5.07 \times 10^4 \;\overset{?}{\lessgtr}\; \left(\frac{2 \times 10^3}{10^{10}} \right)^{1/9} (3.14 \times 10^4) \tag{II-124}$$

$$5.07 \times 10^4 \;\nleq\; 0.565 \times 10^4 \tag{II-125}$$

From this, it is seen that a flat is not an appropriate design for the supporting member. As a result, a flat with a groove of radius, $-\alpha$ (0.030) in., $\alpha \geq 1$, in it must be considered. α can now be determined analytically. Using the formulae for b given for parallel cylinders, we now have:

$$b = 2 \sqrt{ 6.67 \times 2 \times 1.073 \times 10^{-8} \; \frac{(0.03) \; (-0.03\,\alpha)}{0.03 - 0.03\,\alpha} } \;\; in. \tag{II-126}$$

$$b = 1.31 \times 10^{-4} \sqrt{ \frac{\alpha}{\alpha - 1} } \;\; in. \tag{II-127}$$

$$q_o = \frac{2}{\pi} \; \frac{6.67}{1.31 \times 10^4 \sqrt{ \dfrac{\alpha}{\alpha - 1} }} \;\; psi \tag{II-128}$$

$$q_o = 3.24 \times 10^4 \sqrt{ \frac{\alpha - 1}{\alpha} } \;\; psi \tag{II-129}$$

$$\tau_{max} = 3 \left(3.24 \times 10^4 \sqrt{ \frac{\alpha - 1}{\alpha} } \right) \sqrt{ 1/2^2 + 0.15^2 } \;\; psi \tag{II-130}$$

$$\tau_{max} = 5.07 \times 10^4 \sqrt{ \frac{\alpha - 1}{\alpha} } \;\; psi \tag{II-131}$$

N_s and N_r are the same as before. Therefore, substitution into Eqs. (II-101) and (II-103) gives us:

$$5.07 \times 10^4 \sqrt{ \frac{\alpha - 1}{\alpha} } \;\leq\; \left(\frac{2 \times 10^3}{0.625 \times 10^{10}} \right)^{1/9} (3.4 \times 10^4) \tag{II-132}$$

$$5.07 \times 10^4 \sqrt{ \frac{\alpha - 1}{\alpha} } \;\leq\; 0.645 \times 10^4 \tag{II-133}$$

$$5.07 \times 10^4 \sqrt{ \frac{\alpha - 1}{\alpha} } \;\leq\; \left(\frac{2 \times 10^3}{10^{10}} \right)^{1/9} (3.14 \times 10^4) \tag{II-134}$$

$$5.07 \times 10^4 \sqrt{\frac{\alpha - 1}{\alpha}} \quad \leq \quad 0.565 \times 10^4 \qquad \text{(II-135)}$$

The maximum value which α can assume and still have these inequalities obeyed can be obtained by solving the equation obtained by taking the equality expression in Eq. (II-135). The value obtained in this fashion will also satisfy Eq. (II-133) since the right-hand side of (II-135) is less than that of (II-133).

$$5.07 \times 10^4 \sqrt{\frac{\alpha - 1}{\alpha}} \quad = \quad 0.565 \times 10^4 \qquad \text{(II-136)}$$

$$\alpha \quad = \quad 1.01 \qquad \text{(II-137)}$$

This means that the radius of the groove should be no greater than 0.0309 inches. From a practical standpoint, this means that the groove should be the same radius as the rod, i.e., a conforming geometry in order to have zero wear for the desired lifetime.

Layered Material

Example 1

In this example, a magnetic recording system will be evaluated with respect to the wear criterion. The purpose of this evaluation is to determine whether or not there will be zero wear for the desired lifetime. The recording system is composed of a HyMu* 80 head, H_m 270, and a rotating brass disk which has been plated with nickel-cobalt. The geometry of the system is shown in Figure II-6. The contact is a conforming one which can be considered to be a flat on a flat. The head is located 3 inches from the center of the disk and contact between the head and the disk is insured by a total normal load of 1 gram. The disk rotates counter-clockwise about its center and is composed of a nickel-cobalt plating of the order of 20 μ inches on top of a 1/8 inch thick brass disk. This system is required to have zero wear for 10^7 revolutions of the disk.

From the above description of the problem and previous discussion about layered material, it should be realized that since this is a conforming geometry and the thickness of the plating is much smaller than the order of 0.005 inches, it will be necessary to consider the substrate as well as the plated layer, i.e., the brass as well as the nickel-cobalt, and, further, that a microhardness profile of the plating is required.

It will be assumed that the hardness measurements were made and resulted in the graph given in Figure II-7. From this profile, it is seen that the interface between the brass and the nickel-cobalt begins around 20μ inches and extends out to 50 or 60μ inches. In the region less than 20 μ inches, the lowest value of H_m is seen to be approximately 175; this is taken to be the hardness of the plating. Beyond 60 μ inches, the hardness is seen to be approximately H_m 110; this is taken to be the hardness of the substrate. From Figure III-3, it is seen that H_m 175 corresponds to a τ_y of 30 Kpsi, H_m 110, 17 Kpsi.

* Registered Trademark, The Carpenter Steel Company, Reading, Pennsylvania.

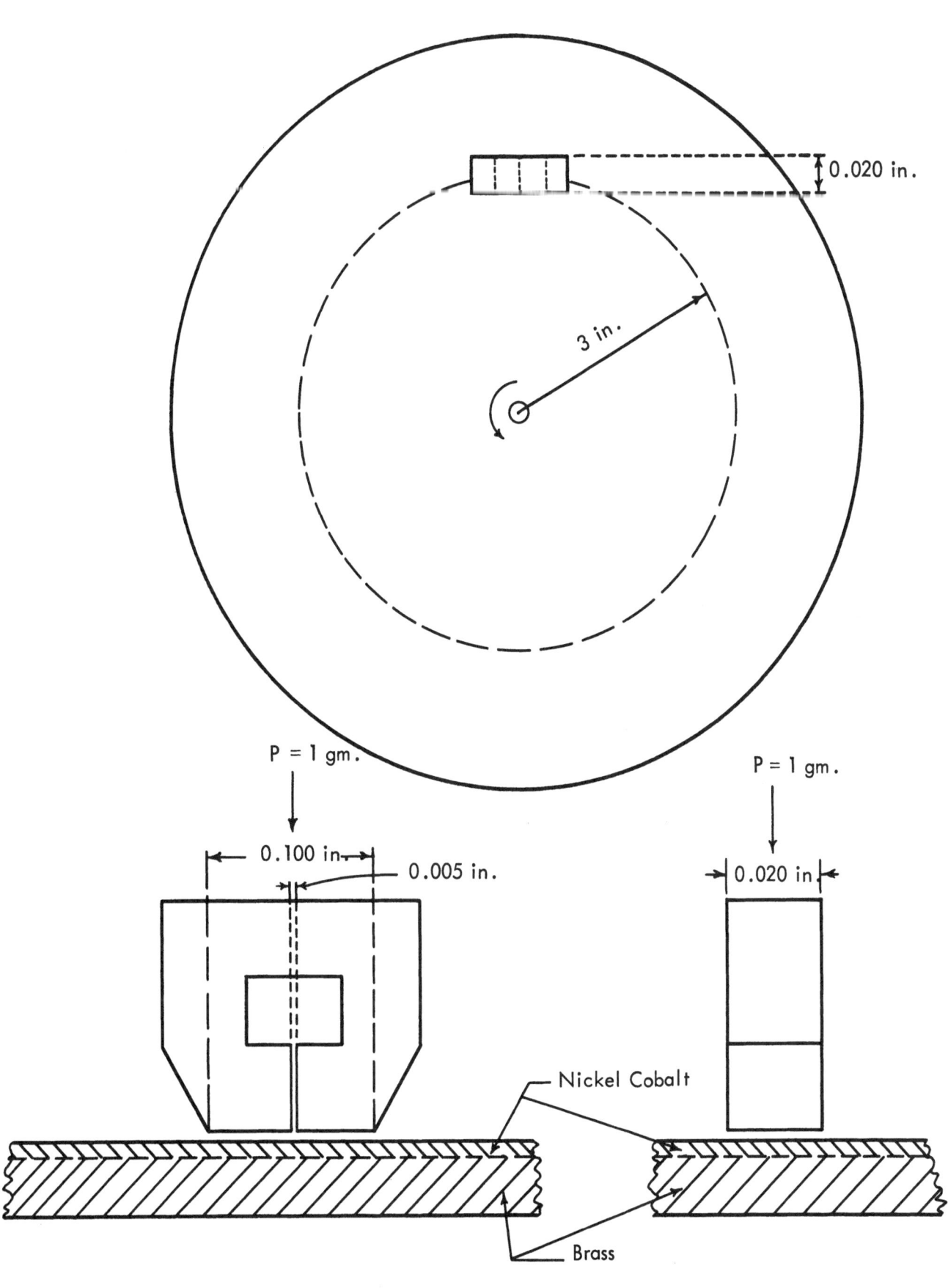

Fig. II-6 – Schematic Of Recording System Considered In Example I
For Layered Materials

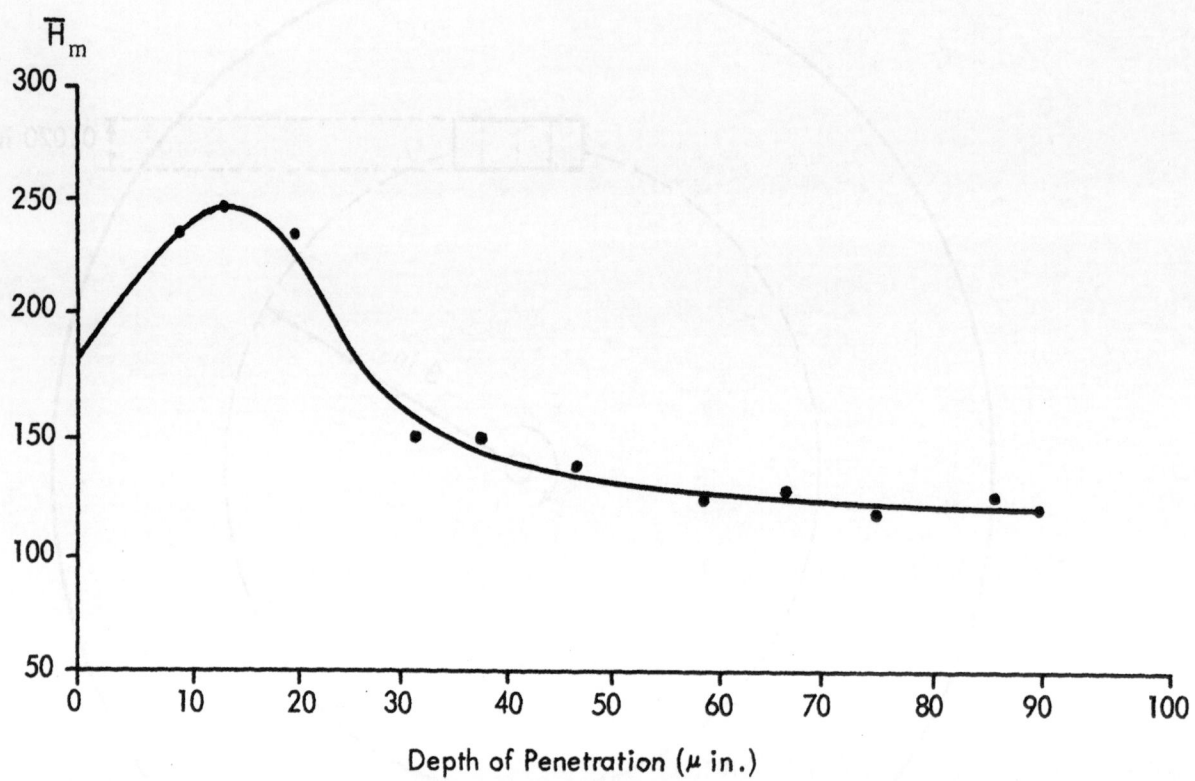

Fig. II-7 – Microhardness Profile Of Plated Disk Considered In Example I
For Layered Materials

The evaluation of this design centers around three inequalities. The design will be satisfactory if the following inequality is satisfied for the head, the plating and the substrate,

$$\tau_{max} \leq \left(\frac{2 \times 10^3}{N} \right)^{1/9} \gamma_R \tau_y \qquad (II-138)$$

when the appropriate values of τ_{max}, N, γ_R and τ_y are used. τ_{max} is the same for all three; it is the maximum shear stress occurring anywhere in the vicinity of the contact. In the determination of N, the head is regarded as the "loaded" member; the nickel-cobalt and the substrate are both regarded as "unloaded" members. γ_R for the head and the nickel-cobalt plating is the same and is the γ_R appropriate for the combination. γ_R for the substrate is 0.54.

The quantity to be determined first will be τ_{max}. τ_{max} is given by

$$\tau_{max} = \frac{KP}{A} \sqrt{1/2^2 + \mu^2} \qquad (II-139)$$

since the geometry is a conforming one. The value of K assumed should be appropriate for sharp edges and corners since the edges usually associated with magnetic heads, especially around the gap, are sharp. Therefore K is taken to be 1000. The value of μ will be estimated since the combination of materials considered here is not listed in the tables of Section III. A value of 1 will be assumed for μ. From Figure II-6, it is seen that A is given approximately by the area of a rectangle of sides 0.100 inches and 0.020 inches. Consequently,

$$\tau_{max} \approx \frac{1000 \times 1 \text{ gm} \times 1 \text{ lb.}/454 \text{ gm}}{.100 \times 0.020} \quad \sqrt{1/2^2 + 1^2} \quad \text{psi} \tag{II-140}$$

$$\tau_{max} \approx 1,240 \text{ psi} \tag{II-141}$$

A single revolution of the disk will be used as the unit pass. It is easily seen that an element on the disk experiences only one loading-unloading cycle in each unit operation. Therefore, n_d for the disk, both for the plating and the substrate is 1/rev. For the head, n_h is given by

$$n_h = \frac{2\pi \text{ (3 in.)}}{0.100 \text{ in.}} = 188/\text{rev} \tag{II-142}$$

Consequently

$$N_d = (1) \times 10^7 \tag{II-143}$$

$$N_d = 10^7 \tag{II-144}$$

$$N_h = (188 \times 10^7) \tag{II-145}$$

$$N_n = 1.88 \times 10^9 \tag{II-146}$$

The values of τ_y for the plating and the substrate were determined previously. The value of τ_y for the head is 55×10^3 psi, corresponding to a H_m of 270 as can be seen from Figure III-3. Since a value of γ_R for HyMu 80 against nickel-cobalt, dry, is not listed in Table III-3, it will be assumed that it is 0.2. Using these values for the τ_y's, γ_R's, etc. and Eq. (II-138), we get the following series of inequalities:

For the head:

$$1.240 \times 10^3 \overset{?}{\lesseqgtr} \left(\frac{2 \times 10^3}{1.88 \times 10^9}\right)^{1/9} \quad (0.20) (5.5 \times 10^4) \tag{II-147}$$

$$1.240 \times 10^3 \leq 2.42 \times 10^3 \tag{II-148}$$

For the plating:

$$1.240 \times 10^3 \overset{?}{\underset{\sim}{\leq}} \left(\frac{2 \times 10^3}{10^7} \right)^{1/9} (0.20)(3.0 \times 10^4) \qquad \text{(II-149)}$$

$$1.240 \times 10^3 \leq 2.35 \times 10^3 \qquad \text{(II-150)}$$

For the substrate:

$$1.240 \times 10^3 \overset{?}{\underset{\sim}{\leq}} \left(\frac{2 \times 10^3}{10^7} \right)^{1/9} (0.54)(1.7 \times 10^4) \qquad \text{(II-151)}$$

$$1.240 \times 10^3 \leq 3.60 \times 10^3 \qquad \text{(II-152)}$$

It is seen that all three inequalities are satisfied. Consequently, the design is satisfactory from a wear standpoint.

Example II

In this example, the design of a reciprocating journal bearing (Figure II-8) will be examined. This bearing is required to take a normal load of 1/4 lb. for 10^6 oscillations of the journal. The journal is an anodized aluminum rod of total radius 0.125 inches. The thickness of the anodized layer is 0.002 to 0.003 inches. The bearing is made of delrin and has a radius of 0.188 inches. The length of the bearing is 0.75 inches. The edges of the bearing are to be well-rounded. Further, there is to be no lubrication and the maximum distance of travel in one direction for the journal is 0.25 inches.

Since the journal is a layered material, it must first be determined whether it is sufficient to treat the anodized layer only or whether it is necessary to consider both the anodized layer and the aluminum substrate. In this case, since this geometry is not a conforming one, this is determined by means of a comparison of 2b and the thickness of the anodized layer. If the thickness is of the order of or greater than 2b, only the anodized layer need be considered. If not, both the anodized layer and the substrate must be considered. Since E and ν of the anodized layer and the aluminum substrate are taken to be the same (Table III-2), 2b is given as

$$2b = 4 \sqrt{\frac{P' R_1 R_2 (k_1 + k_2)}{R_1 + R_2}} \qquad \text{(II-153)}$$

where the indices 1 and 2 will refer to the journal and the bearing, respectively, and k_2 and k_1 are computed using the elastic constants for delrin and aluminum, respectively (Table III-2). For the problem considered then

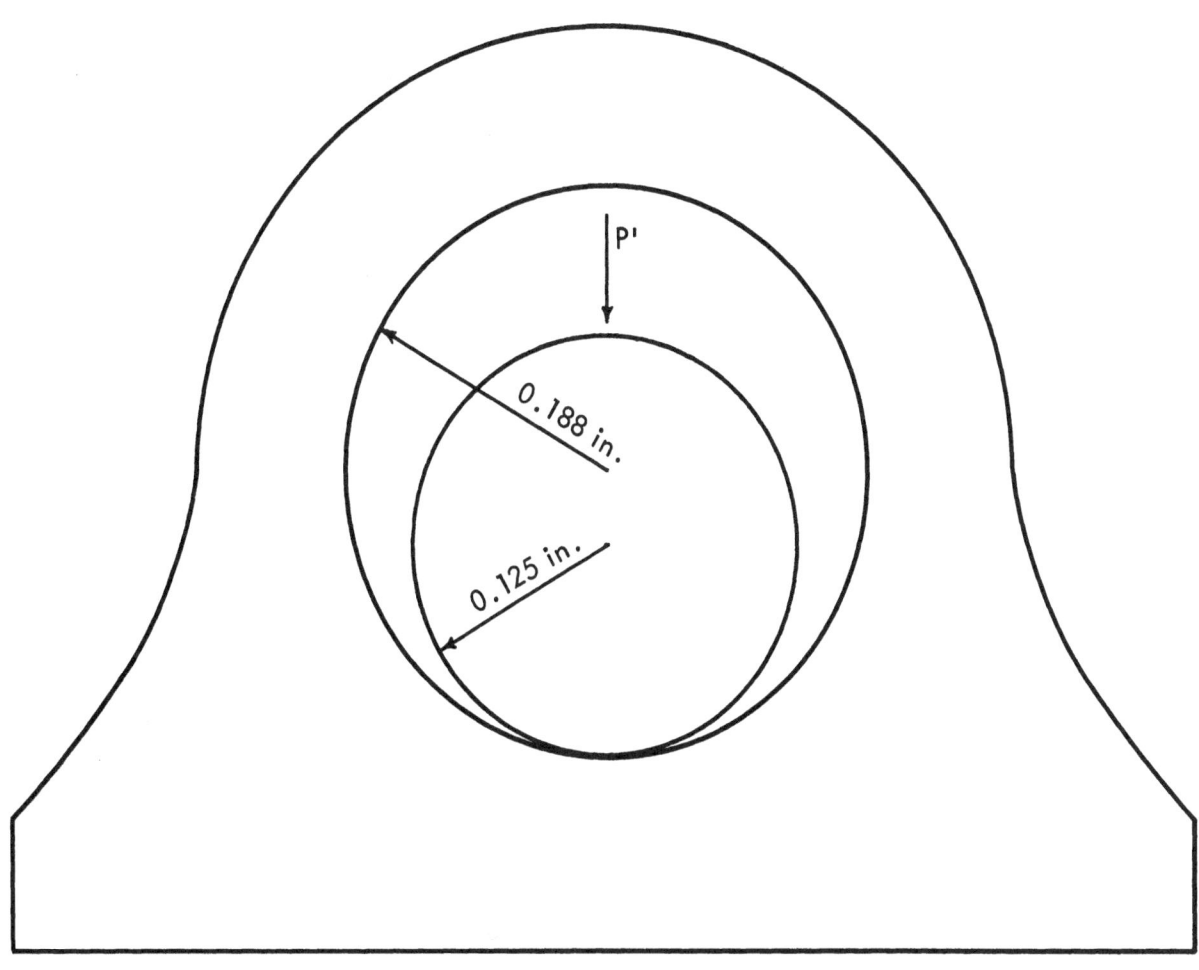

Fig. II-8 - Schematic Of The Mechanism Considered In Example II
For Layered Material

$$P' = \frac{0.25 \text{ lb.}}{0.75 \text{ in.}} \tag{II-154}$$

$$P' = 0.333 \; \frac{\text{lb.}}{\text{in.}} \tag{II-155}$$

$$R_1 = 0.125 \text{ in.} \tag{II-156}$$

$$R_2 = -0.188 \text{ in.} \tag{II-157}$$

$$k_1 = \frac{1 - (0.33)^2}{\pi \times 10^7} \quad (\text{psi})^{-1} \tag{II-158}$$

$$k_1 = 0.284 \times 10^{-7} \quad (\text{psi})^{-1} \tag{II-159}$$

$$k_2 = \frac{1 - (0.35)^2}{\pi \, (4.75 \times 10^5)} \quad (\text{psi})^{-1} \tag{II-160}$$

$$k_2 = 0.590 \times 10^{-6} \quad (\text{psi})^{-1} \tag{II-161}$$

$$2b = 4 \sqrt{\frac{0.333 \times (0.125)\,(-0.188)\,(0.284 \times 10^{-7} + 0.590 \times 10^{-6})}{0.125 - 0.188}} \quad \text{in.} \tag{II-162}$$

$$2b = 0.00112 \text{ in.} \tag{II-163}$$

Consequently, since the anodized layer is 0.002 – 0.003 inches thick, only the anodized layer will be considered.

The next step will be to determine the hardness of the anodized layer. It will be assumed that the microhardness measurements of the anodized layer resulted in a value of H_m 258, which corresponds to a value of 50×10^3 psi for τ_y (Figure III-3).

If this mechanism is to have zero wear, then the following inequality must be satisfied in both the delrin bearing and the anodized layer,

$$\tau_{max} \leq \left(\frac{2 \times 10^3}{N} \right)^{1/9} \gamma_R \; \tau_y \tag{II-164}$$

To determine τ_{max}, q_o must first be determined

$$q_o = \frac{2}{\pi} \frac{P'}{b} \qquad\qquad (II-165)$$

$$q_o = \frac{2}{\pi} \frac{0.333}{0.00112} \quad psi \qquad\qquad (II-166)$$

$$q_o = 380 \quad psi \qquad\qquad (II-167)$$

τ_{max} is given by

$$\tau_{max} = K \, q_o \, \sqrt{1/2^2 + \mu^2} \qquad\qquad (II-168)$$

Since the edges are specified to be well-rounded, K will be taken to be 3. The value of μ for this particular combination of materials is not listed in the tables of Section III. However, a value for the combination of 302-delrin, dry, is given as 0.36 and for anodized aluminum and 52100, dry, as 0.16. Therefore, as an estimate for the combination given here a value of 0.5 will be taken, in lieu of an experimental determination. τ_{max} in this case is

$$\tau_{max} = 3 \times 380 \, \sqrt{1/2^2 + 1/2^2} \quad psi \qquad\qquad (II-169)$$

$$\tau_{max} = 801 \quad psi \qquad\qquad (II-170)$$

The next step will be to determine the number of passes experienced by the journal, N_j, and the bearing, N_b, in 10^6 oscillations. The journal is the "unloaded" member in this mechanism. However, since the length of the bearing is 0.75 inches and the maximum distance of travel in one direction is only 0.25 inches, the contact zone is not fully unloaded. Consequently, the unit operation is taken to be a single oscillation. The total sliding distance, S, in a single oscillation is 2×0.25 inches or 0.50 inches. Therefore

$$n_j = \frac{S}{W} \qquad\qquad (II-171)$$

$$n_j = \frac{0.50}{0.75} \, /osc \qquad\qquad (II-172)$$

$$n_j = 0.667/\text{osc} \tag{II-173}$$

$$n_b = 2/\text{osc} \tag{II-174}$$

and

$$N_j = 0.667 \times 10^6 \tag{II-175}$$

$$N_b = 2 \times 10^6 \tag{II-176}$$

As was mentioned previously, this particular combination of materials is not contained in Table III-3. Therefore γ_R will be estimated. In Table III-3, it will be noted that for all combinations in which a plastic appears and in which anodized aluminum appears, $\gamma_R = 0.54$. Consequently, it will be estimated that $\gamma_R = 0.54$. From Table III-4, it is seen that the τ_y for delrin is 1.235×10^3 lb/in.

Using these values, Eq. (II-163) when applied to the bearing becomes

$$801 \overset{?}{\leq} \left(\frac{2 \times 10^3}{2 \times 10^6}\right)^{1/9} (0.54)(1.235 \times 10^3) \tag{II-177}$$

$$801 \nleq 313 \tag{II-178}$$

For the journal

$$801 \overset{?}{\leq} \left(\frac{2 \times 10^3}{6.68 \times 10^5}\right)^{1/9} (0.54)(5.0 \times 10^4) \tag{II-179}$$

$$801 \leq 1.416 \times 10^4 \tag{II-180}$$

It is seen from Eqs. (II-178) and (II-180) that as far as the journal is concerned the design is satisfactory, but that the wear should be expected to occur on the bearing. Consequently, the overall design is not satisfactory.

B. DESIGN PROCEDURE FOR NON-ZERO WEAR

The design procedure for non-zero wear is similar to that for zero wear in the sense that analytical expressions are used to determine suitable combinations of design parameters. For non-zero wear, the procedure is somewhat more complicated than for the zero wear case since there is no simple algebraic expression available for relating lifetime and design parameters for the general case, as there is for zero wear. The expressions available for the general case of non-zero wear are differential equations, which must be integrated for each individual problem to obtain the algebraic expression relating wear to load, geometry, number of operations, and material properties. These expressions are then used to determine the suitable combination of design parameters in a fashion similar to that done in the zero wear analysis. The usual criterion for this determination in the case of non-zero wear is that the wear stay under some maximum allowable limit for the number of operations desired.

It will be assumed that the reader is acquainted with the design procedure for zero wear and consequently the discussions concerning τ_{max}, τ_y, γ_R and N will not be repeated here. However, the formulation and integration of the differential equations and the determination of the dependencies of wear on load, geometry and material properties will be discussed in detail below.

The development of the necessary expressions for the determination of suitable combinations of design parameters usually involves three main steps.

The first step is to determine of which of the following differential equations is appropriate for the mechanism considered.

$$dQ = C'dN \tag{II-181}$$

which is appropriate for Type A wear, or

$$dQ = C'' (\tau_{max} \ W)^{9/2} \ dN + \frac{9}{2} \ \frac{Q}{(\tau_{max} \ W)} \ d(\tau_{max} \ W) \tag{II-182}$$

which is appropriate for Type B wear.

The second step involves the integration of the appropriate differential equations. This step results in a relationship between Q and L which, in addition, usually involves parameters which depend on load, geometry and material properties. The third step is the determination of the dependencies of these parameters on these quantities.

At the conclusion of these three steps, expressions are then available which can be used in the same fashion as Eq. (II-1) to determine whether a given set of design parameters is satisfactory, or what values certain of these parameters must assume so that the wear will be acceptable.

It is significant to point out that some knowledge of the relationship between wear and the loads, materials, geometries and lubricants which are involved in the mechanism must be available in order to perform the first two steps. The first step requires such knowledge since, in order to determine the appropriate equation, it must be known whether the wear can be described as Type A or B. In making such a determination, a knowledge of the materials and lubricant to be used is essential. For example, if the materials of the mechanism were to be aluminum and steel and there was to be no lubrication, Type A wear would be expected to occur if the stress levels were high enough; while if a lubricant was to be used, it would be expected that Type B wear would occur. In the first case, Eq. (II-181) would be used to describe the system; while in the second, Eq. (II-182) would be used.

The second step requires such information since if Eq. (II-182) is chosen, some expressions must be developed for Q and τ_{max} W in such a manner that the equation can be integrated. This can be done by expressing Q and (τ_{max} W) in terms of a common parameter, which can be used as an indicator of how much wear has occurred. This is usually done through a knowledge of the geometry of the wear scar and the geometry of the members of the mechanism in the worn state. The geometry of the wear scar, and likewise the geometry of the members in the worn state, can be inferred in certain cases if certain properties of the materials and the original geometries to be used are known. For example, a hard steel cylinder, sliding in a direction parallel to its axis, and against a soft steel flat would be expected to wear a trough of circular cross-section in the flat. Consequently, in the worn state, the geometry of the contact would be that of a cylinder in a circular trough, instead of the original cylinder — plane geometry. To illustrate the first two steps of the technique, this example of a cylinder on a plane will now be considered in more detail.

Consider a cylinder of length ℓ and radius r, sliding against a flat surface (Figure II-9(a)). The motion is a back-and-forth motion in a direction parallel to the axis of the cylinder. The overall length of travel in one direction is S, $S > \ell$. The material of the cylinder is a hardened steel; the flat is aluminum. The problem is to determine how wear will progress after the conditions for zero wear are violated. Two conditions will be considered, the first when there is a lubricant used, and, the second when there is none used and the stress levels are high in comparison to τ_y of the aluminum. Under the first condition it would be expected that transfer would not predominate and that Type B wear would occur. Consequently, Eq. (II-182) applies. Under the second, Eq. (II-181) is the appropriate equation since Type A wear would be expected to occur. In either case, the scar would be expected to be a groove, of the same radius as the cylinder, in the flat. Figure II-9(b) gives an idealized cross-section of the scar with a cross-sectional view of the cylinder superimposed.

For the first condition, i.e., when a lubricant is used, it is necessary to develop expressions for Q and τ_{max} W in terms of a common parameter so that Eq. (II-182) may be integrated. This may be done in the following manner in which both of the quantities will be expressed in terms of the width of the scar, T.

If the depth of the scar, h, is small in comparison to the radius of the cylinder, the scar may be approximated by a triangle. Therefore,

$$Q \approx \frac{1}{2} hT \qquad (II-183)$$

(If h is larger, Eq. (II-183) would become more complex). From the geometry given in Fig. II-10(b), it is seen that

$$h \approx \frac{T^2}{8r} \qquad (II-184)$$

since

$$(T/2)^2 + (r-h)^2 = r^2 \qquad (II-185)$$

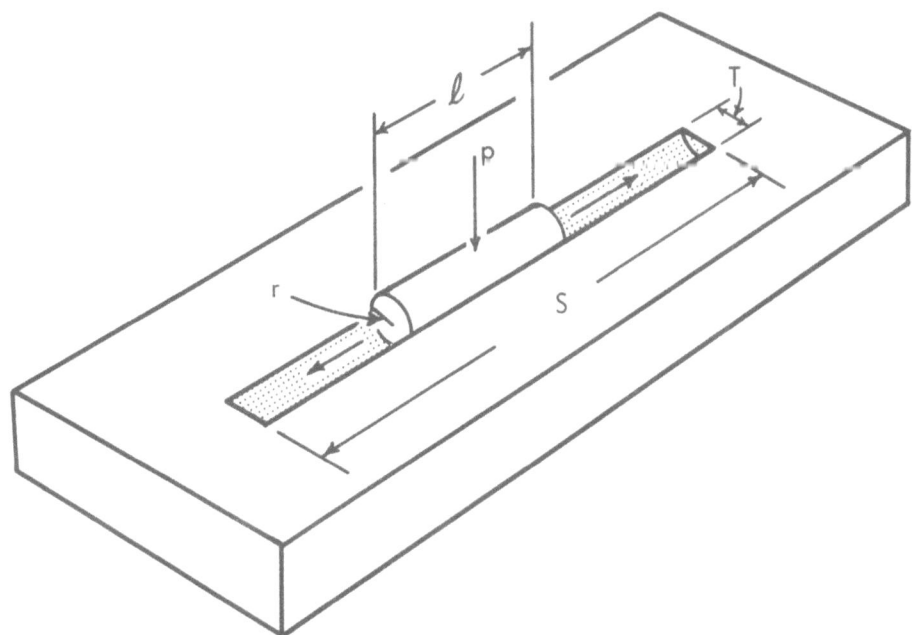

Fig. II-9 (a) - Overall View

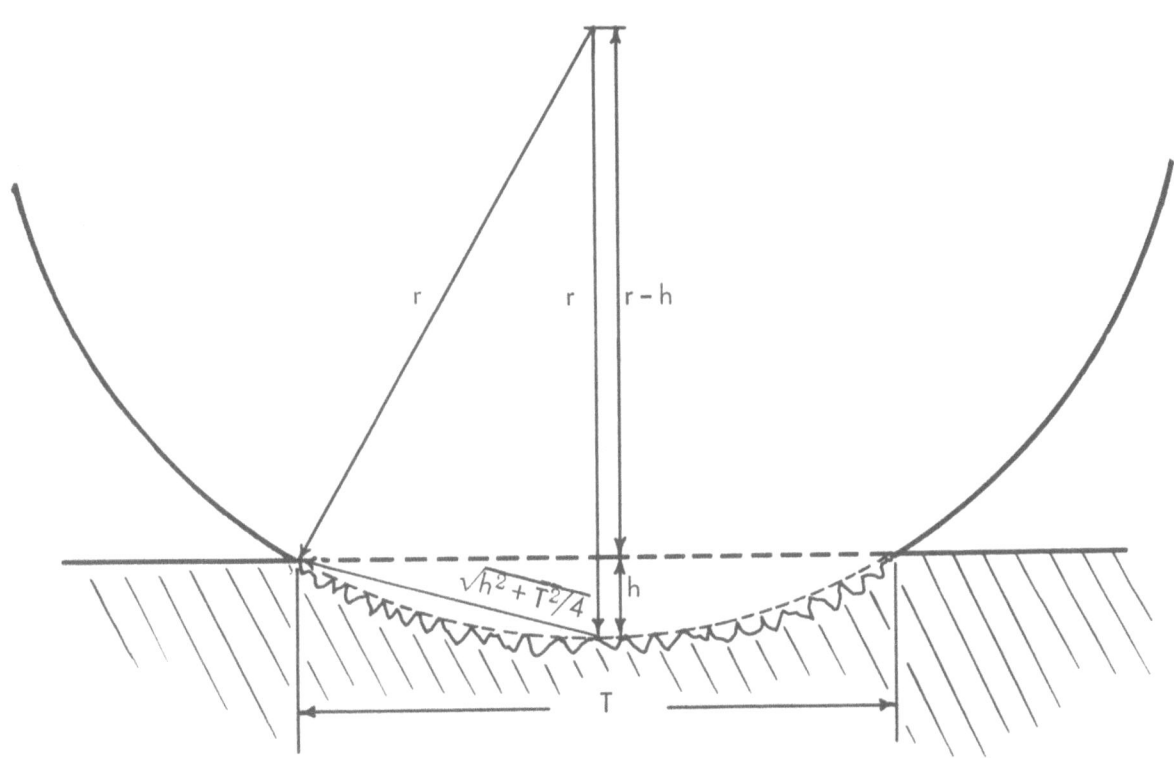

Fig. II-9 (b) - Cross Sectional View Of Wear Scar In The Flat

Fig. II-9 - Wear Of A Soft Aluminum Flat By A Hard Steel Cylinder

$$\frac{T^2}{4} + r^2 - 2rh + h^2 = r^2 \tag{II-186}$$

and

$$h^2 \ll 2rh \tag{II-187}$$

Therefore

$$Q \approx \frac{T^3}{16r} \tag{II-188}$$

Now since the contact is a conforming one

$$\tau_{max} = K \frac{P}{A} \sqrt{1/2^2 + \mu^2} \tag{II-189}$$

where

$$A = TW \tag{II-190}$$

$$W = \ell \tag{II-191}$$

Therefore

$$\tau_{max} = \frac{KP}{T\ell} \sqrt{1/2^2 + \mu^2} \tag{II-192}$$

Consequently

$$\tau_{max} W = \frac{KP\sqrt{1/2^2 + \mu^2}}{T} \tag{II-193}$$

Eqs. (II-188) and (II-193) allow Eq. (II-182) to be integrated since they express Q and τ_{max} W, respectively, in terms of a single variable, T. From Eqs. (II-188) and (II-193), we have

$$dQ \approx \frac{3T^2}{16r} \, dT \tag{II-194}$$

$$d(\tau_{max} W) = - \frac{KP\sqrt{1/2^2 + \mu^2}}{T^2} \, dT \tag{II-195}$$

Before Eq. (II-182) can be integrated, an additional point must be considered, namely the variation of N with Q. Since the size of the contact changes with wear, it is possible to have a change in the number of passes experienced by a member in a unit operation. This is accounted for by taking the ratio of dN to dL to be n, i.e.,

$$dN = ndL \tag{II-196}$$

For the flat in this problem n is constant and equal to 1, if a single stroke is taken as the unit operation. Therefore,

$$dN = dL \tag{II-197}$$

Substituting Eq. (II-188), (II-193), (II-194), (II-195) and (II-196) into Eq. (II-182), Eq. (II-198) is obtained

$$\frac{3T^2}{16r} dT = C'' K^{9/2} P^{9/2} (1/2^2 + \mu^2)^{9/4} \frac{dL}{T^{9/2}}$$

$$+ \frac{9}{2} \frac{(T^3/16r)}{\left(\frac{KP\sqrt{1/2^2 + \mu^2}}{T}\right)} \left(-\frac{KP\sqrt{1/2^2 + \mu^2}}{T^2} dT\right) \tag{II-198}$$

Then proceeding as follows, Eq. (II-202) is obtained:

$$\frac{3T^2}{16r} dT = C'' K^{9/2} P^{9/2} (1/2^2 + \mu^2)^{9/4} \frac{dL}{T^{9/2}} - \frac{9}{32r} T^2 dT \tag{II-199}$$

$$\frac{1}{r}\left(\frac{6}{32} + \frac{9}{32}\right) T^2 dT = C'' K^{9/2} P^{9/2} (1/2^2 + \mu^2)^{9/4} \frac{dL}{T^{9/2}} \tag{II-200}$$

$$T^{13/2} dT = \frac{32}{15} C'' K^{9/2} P^{9/2} (1/2^2 + \mu^2)^{9/4} r dL \tag{II-201}$$

$$T^{15/2} = 16C'' K^{9/2} P^{9/2} (1/2^2 + \mu^2)^{9/4} rL + C_2 \tag{II-202}$$

where C_2 is a constant of integration. Using Eq. (II-188), Eq. (II-202) can be put in the following form,

$$Q^{5/2} = \frac{C'' K^{9/2} P^{9/2} (1/2^2 + \mu^2)^{9/4} L}{64 r^{3/2}} + C_2 \tag{II-203}$$

$$Q^{5/2} = C_1 L + C_2 \tag{II-204}$$

where

$$C_1 = \frac{C'' \, K^{9/2} \, P^{9/2} \, (1/2^2 + \mu^2)^{9/4}}{64 \, r^{3/2}} \tag{II-205}$$

Eq. (II-204) gives the dependency of Q on L. The dependencies of Q on other parameters of the system are contained in the quantities C_1 and C_2 of Eq. (II-204). As was mentioned earlier, the determination of these dependencies constitutes the third step. Before these parameters are discussed in more detail, it should be realized that this equation (Eq. (II-204)) implicitly places a limit on the variation allowed in certain of these parameters. In using this equation, these parameters cannot be allowed to assume values in a range for which the assumptions made in obtaining Eq. (II-204) are invalid. For example, if the dependency of Q on types of lubrication is of interest, Eq. (II-204) cannot be used in the investigation of lubricants which permit large amounts of transfer to occur; or if the dependency of Q on material composition of the cylinder is of interest, Eq. (II-204) cannot be directly applied to a case in which the cylinder is softer than the flat.

One way of determining the C's in Eq. (II-204) is to perform a series of controlled experiments, in which Q is determined for two different numbers of operations for the various values and combinations of parameters of interest. These values of Q and L enable C_1 and C_2 to be determined. However, the determination of the C's and their dependencies on certain of the parameters need not be done entirely on an experimental basis. There are certain analytical aids which can be used.

One analytical approach is for a case in which there is a period of at least 2000 passes of what may be called zero wear before the wear has progressed to beyond the surface finish. In such a case, C_1 and C_2 can be estimated analytically. This is done by taking C_2 to be zero and determining C_1 from the model for zero wear. C_1 is determined by first finding the maximum number of operations, L_1, for which there will be zero wear for the load, geometry, etc. of interest.

L_1 is given by

$$L_1 = \frac{2 \times 10^3}{n} \left(\frac{\gamma_R \, \tau_y}{\tau_{max}} \right)^9 \tag{II-206}$$

where τ_{max} is the maximum shear stress computed using the unworn geometry.

The wear, Q_1, produced and accumulated during this number of passes is taken to be a scar of the same profile assumed in deriving Eq. (II-204) or a similar equation, and of a depth equal to one-half the peak-to-peak surface roughness of the material.

In the particular problem considered, if we take the surface roughness, peak-to-peak, to be 30 μ in., then

$$Q \approx \sqrt{2r} \; (15 \times 10^{-6} \text{ in.})^{3/2} \tag{II-207}$$

as can be seen from Eqs. (II-183) and (II-184).

Q_1 and L_1 would then be substituted into Eq. (II-204), or a similar one, to determine C_1.

Another analytical aid which is useful for other cases is that it has been found that C" in Eq. (II-182) or (II-203) is approximately independent of P. Consequently the dependency of C_1 on P is known.

Under the second set of conditions, i.e., when a lubricant is not used, and the stress levels are high enough, Type A wear would occur. As a result, Eq. (II-181) would be the appropriate one to use and upon integration this would result in

$$Q = C_1' L + C_2' \tag{II-208}$$

where L has been introduced in the same manner as before. C_1' and C_2' must be determined experimentally.

It is important to realize that even though, strictly speaking, L in Eqs. (II-204) and (II-208) refers to number of operations after the geometry of the contact has been worn to the state assumed in the formulation of the problem, it can be treated as an absolute number of operations, i.e., counted from the very first operation of the device, as long as the constants are determined in the manner discussed. Used in such a fashion, such expressions, when used in conjunction with a knowledge of the conditions required for zero wear (which are obtained from the model for zero wear), afford a complete profile of the wear phenomenon associated with a given mechanism in terms of load, geometry, material properties and number of operations.

The example discussed above concerns itself with a relatively simple case of wear; nevertheless it demonstrates the major steps required in developing the expressions needed for design. In general, of course, the developing of such expressions may be more difficult than for the case considered. In particular, the example considered assumed wear on only one member. Quite often wear will occur on both members of a mechanism. In such a case, the application of the model would result in a pair of simultaneous differential equations, each describing the wear on one member, and the ensuing difficulties of integration. Quite often in such cases, it is possible to obtain significant design information by certain approximations and idealizations. One such approximation would be to estimate the wear of one member and its dependencies on design parameters by assuming the wear on the other member to be negligible. Such approaches usually result in less accurate specifications than a more precise treatment; however they are quite useful in estimating suitable ranges of parameters.

It is seen from the above that the analysis of a mechanism for non-zero wear is considerably more complicated than that for zero wear. However, the above discussion has indicated the general outline of how such an analysis is performed. An example of such an analysis is now given.

Example

The problem which will be considered is the evaluation of a design modification of the mechanism considered in Example 1 for a homogeneous material. The mechanism considered in that example was composed of a hemispherically ended rod riding on the circumference of a rotating cylinder. The radius of the hemisphere was 0.5 inches; the cylinder, 2.00 inches. The material for the hemisphere was 52100, hardened to a microhardness of H_m 746, and 8620, H_m 216 for the cylinder. The lubricant was Oil B and the normal load was 1 oz. In that example, it was determined that there would be zero wear for at least 10^6 revolutions of the cylinder. The modification which will now be considered is an increase in the normal load to 2 oz. and the use of "soft" 52100, H_m 220, for the hemisphere instead of "hard" 52100; H_m 746. It is required that this mechanism be able to operate for 10^6 revolutions with the maximum amount of total wear in the direction along the axis of the rod to be less than or equal to 0.005 inches.

The first step in the examination of this modified design will be in terms of the zero wear criterion. If the criterion for zero wear is satisfied, the design is satisfactory. If not, a further analysis, based on the model for non-zero wear, is required.

The zero wear evaluation concerns itself with the determination of whether or not the following inequality is satisfied by each member

$$\tau_{max} \leq \left(\frac{2 \times 10^3}{N} \right)^{1/9} \gamma_R \, \tau_y \tag{II-209}$$

Since only the load and hardness of the 52100 have been changed from that considered previously, determination of several of the quantities needed for a zero-wear evaluation can be simplified. In particular, since the composition of "soft" and "hard" 52100 is the same, γ_R is the same, 0.54. Consequently the right-hand side of the inequality when applied to the cylinder is numerically the same as before; the left side will change because of the increase in load.

Since the stress is proportional to $P^{1/3}$ for the general Hertz case, and likewise "a" and "b", the new values of τ_{max}, "a" and "b" can be obtained by multiplying the previously obtained values by the cube root of the ratio of the new load to the old. Therefore

$$\tau_{max} = 7.95 \times 10^3 \quad \sqrt[3]{2/1} \text{ psi} \tag{II-210}$$

$$\tau_{max} = 1.00 \times 10^4 \text{ psi} \tag{II-211}$$

$$a = 0.116 \times 10^2 \quad \sqrt[3]{2/1} \text{ in.} \tag{II-212}$$

$$a = 0.146 \times 10^{-2} \text{ in.} \tag{II-213}$$

$$b = 0.100 \times 10^{-2} \ \sqrt[3]{2/1} \ \text{in}. \tag{II-214}$$

$$a = 0.126 \times 10^{-2} \ \text{in}. \tag{II-215}$$

Since b has changed because of the load, N_s and n_s have also changed for 10^6 revolutions. The new N_s is given by the product of the old N_s times the value of the previous b divided by the current b, i.e.,

$$N_s = 6.28 \times 10^9 \ \times \ \frac{0.100 \times 10^{-2}}{0.126 \times 10^{-2}} \tag{II-216}$$

$$N_s = 4.98 \times 10^9$$

Since the hardness of the 52100 is changed, the value of τ_y for the sphere is also changed. The value of τ_y corresponding to H_m 220 is seen from Figure III-3 to be 40×10^3 psi.

For the cylinder, Eq. (II-209) becomes

$$1.00 \times 10^4 \leq 1.09 \times 10^4 \tag{II-218}$$

Consequently, as far as the cylinder is concerned, the modifications are satisfactory. For the sphere, Eq. (II-209) becomes

$$1.00 \times 10^4 \ \overset{?}{\lessgtr} \ \left(\frac{2 \times 10^3}{4.98 \times 10^9} \right)^{1/9} \ (0.54) \ (40 \times 10^3) \tag{II-219}$$

$$1.00 \times 10^4 \ \not< \ 4.28 \times 10^3 \tag{II-220}$$

Consequently, the modified design does not insure zero wear for the sphere. However, since the mechanism will perform properly, even if some wear does occur, the design will now be examined, in terms of the model for non-zero wear, to see whether the wear produced will or will not exceed the allowable limit.

The first step in the analysis is to formulate the differential equations describing the wear process. It is seen that since the zero-wear criterion for the cylinder is satisfied, only one equation need be formulated — the one describing the wear on the sphere. The differential equation to be used is Eq. (II-182),

$$dQ = C'' \, (\tau_{max} \, W)^{9/2} \, dN + \frac{9}{2} \, \frac{Q}{(\tau_{max} \, W)} \, d \, (\tau_{max} \, W) \tag{II-221}$$

since Type A wear would not be expected to occur between steels, especially under lubricated conditions and with stresses considerably below the yield points of the materials.

Having determined which is the appropriate differential equation to describe the wear of the sphere, the next step is to determine the geometry of the wear scar and from this certain properties of the contact in the worn state, since τ_{max}, W, Q, and N are intimately related to these.

It can be inferred from the calculations concerning the zero wear criterion for this mechanism that the cylinder will not experience any "non-zero" wear during the lifetime specified, but will wear a conforming groove into the sphere (Figure II-10).

The first property of the contact to be determined will be its projected area on a plane perpendicular to the line of action of the force. This area will be later used to determine τ_{max}.

The shape of this projected area can be obtained from the following considerations. Consider the contact area to be sliced by a series of planes perpendicular to the line of action of the force. For each plane, determine the curve formed by the intersection of that plane and the sphere as well as the curve formed by the intersection of that plane with the cylinder. The curves for the spheres are circles; those for the cylinder are parallel straight lines. The points at which these two curves intersect, i.e. the intersection of the circle and the straight lines corresponding to a given plane, are points on the boundary of the desired projected area of contact. Figure II-11 shows a construction demonstrating these considerations. It can be seen in Figure II-11 that this area, A, can be approximated by an ellipse, and consequently

$$A \approx \pi \alpha \beta \tag{II-222}$$

where α and β are the semi-axes of the ellipse, α being the semi-axis in a direction parallel to the axis of the cylinder.

The next two quantities to be determined are Q and W. It is convenient to express these two quantities, as well as the α and β defined above, in terms of a single wear index. Any quantity which varies with wear, such as α and β, would be suitable. However, since the limit for wear is specified in terms of maximum wear in a direction along the axis of the rod, it would be convenient to use this quantity, i.e., maximum wear in this direction, as the index. This quantity is the "h" indicated in Figure II-11.

From Figure II-11 (a), the relationship between α and h and Q and h can easily be determined. From this figure, it can be seen that

$$r^2 = (r - h)^2 + \alpha^2 \tag{II-223}$$

Since h << r, for the problem and range of wear considered

$$\alpha \approx \sqrt{2rh} \tag{II-224}$$

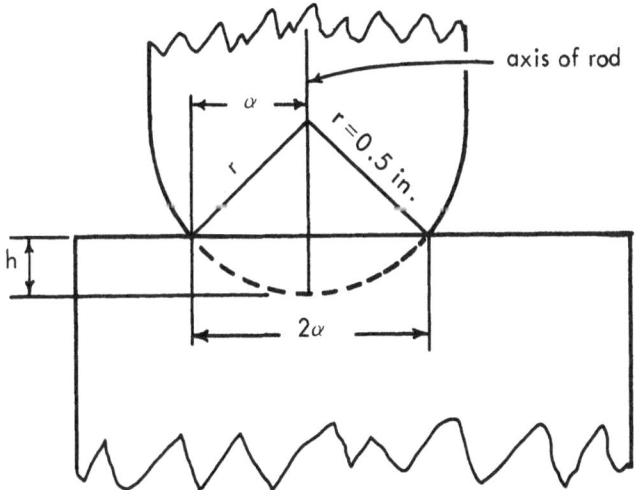

Fig II - 10 (a) - View In Plane Containing Axis Of Rotation And Axis Of Rod

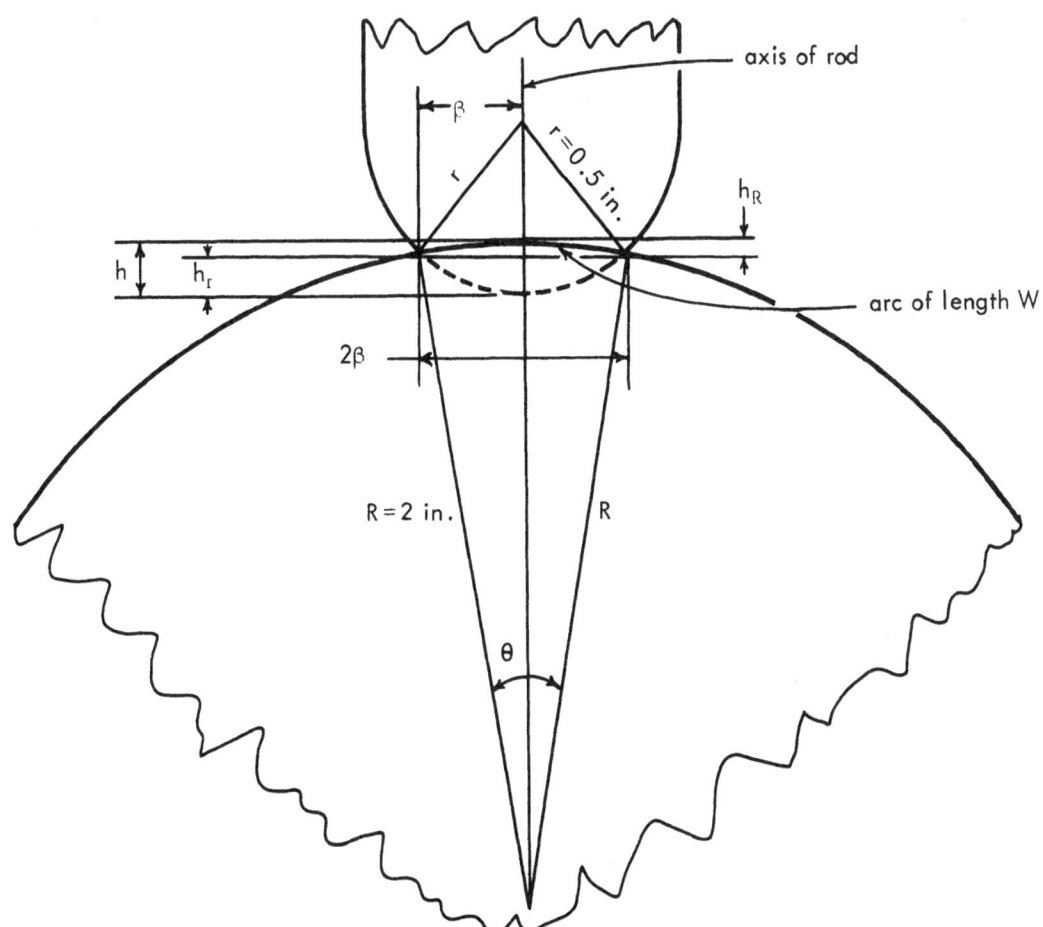

Fig II - 10 (b) - View In Plane Perpendicular To Axis Of Rotation

Fig. II - 10 - Schematic Of Mechanism Considered In Example For Non-Zero Wear In The Wear State

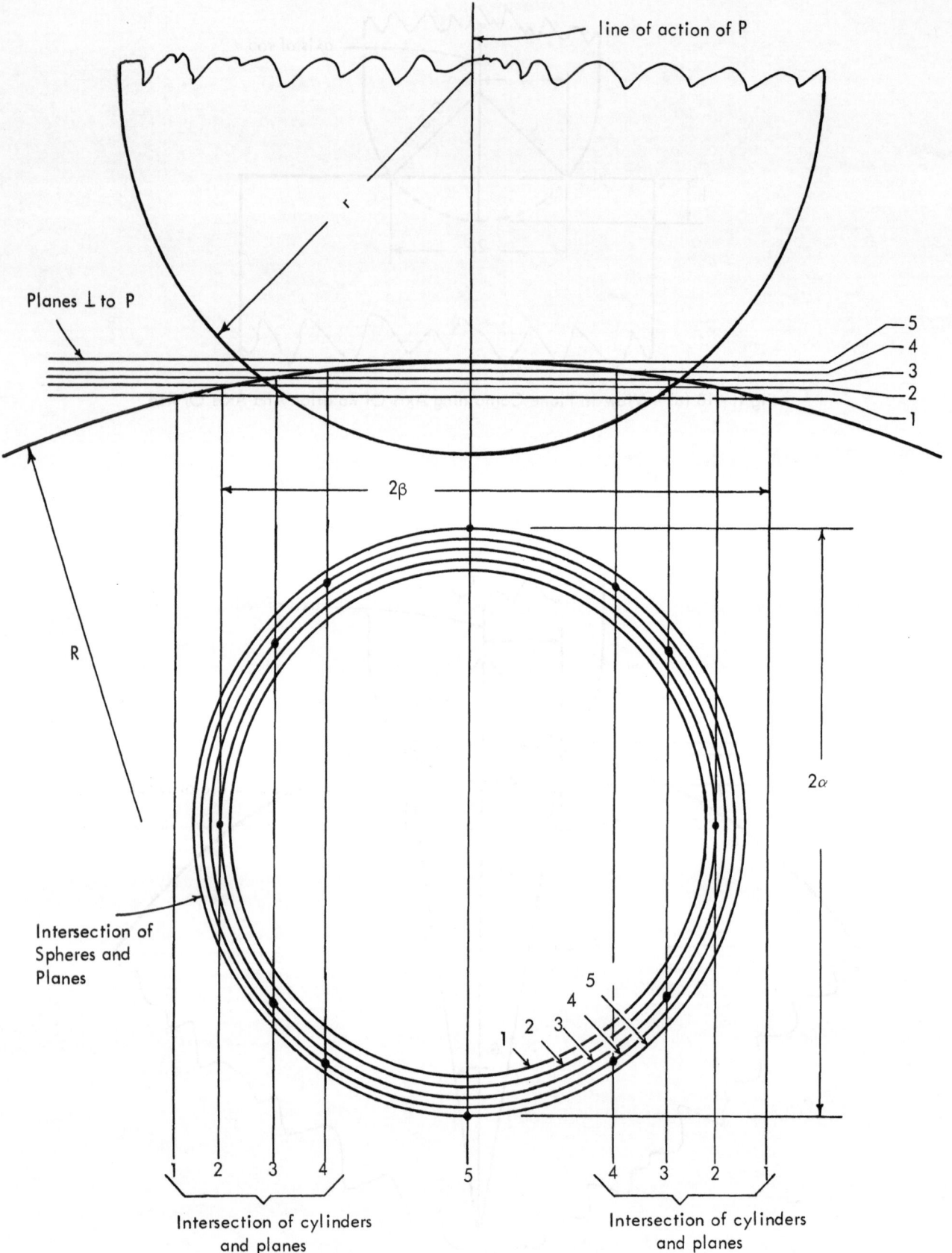

Fig. II-11 - Diagram Of Technique For Determining Project Area In Example For Non-Zero Wear

Since $h \ll r$, the cross-sectional area of the scar taken in a plane perpendicular to the direction of motion can be approximated by a triangle. Therefore,

$$Q \approx \alpha h \tag{II-225}$$

Substituting Eq. (II-224)

$$Q \approx \sqrt{2rh^3} \tag{}$$

The relationship between β and h and W and h can be obtained from the view given in Figure II-10(b). It is seen from that figure that

$$h = h_R + h_r \tag{II-227}$$

and that

$$R^2 = (R - h_R)^2 + \beta^2 \tag{II-228}$$

$$r^2 = (r - h_r)^2 + \beta^2 \tag{II-229}$$

Again, since $h \ll r < R$,

$$h_R \approx \frac{\beta^2}{2R} \tag{II-230}$$

$$h_r \approx \frac{\beta^2}{2r} \tag{II-231}$$

Therefore, from Eq. (II-227)

$$h \approx \frac{\beta^2}{2} \left(\frac{1}{R} + \frac{1}{r} \right) \tag{II-232}$$

$$\beta \approx \sqrt{\frac{2rRh}{R+r}} \tag{II-233}$$

W, as is seen in Figure II-10 is the arc, of a circle of radius R, whose chord is 2β. Consequently,

$$W = \theta R \tag{II-234}$$

where

$$\theta = \sin^{-1} \frac{\beta}{R} \qquad \text{(II-235)}$$

Eq. (II-235) and therefore Eq. (II-234) can be simplified if one considers the magnitudes of β and R in the problem considered. If the values for r and R are substituted in Eq. (II-233), we get

$$\beta \approx 0.885 \sqrt{h} \quad \text{in.}^{1/2} \qquad \text{(II-236)}$$

Since the range of h of interest is from 0 to 0.005 inches, we see that

$$0 \leq \beta \leq 0.0625 \text{ in.} \qquad \text{(II-237)}$$

Consequently,

$$0 \leq \sin \theta \leq \frac{0.0625}{2} \qquad \text{(II-238)}$$

$$0 \leq \sin \theta \leq 0.0313 \qquad \text{(II-239)}$$

This is within the range where

$$\sin \theta \approx \theta \qquad \text{(II-240)}$$

Therefore, Eq. (II-234) can be written as

$$W \approx \beta \qquad \text{(II-241)}$$

Substituting for β from Eq. (II-233), we get

$$W \approx \sqrt{\frac{2rRh}{R+r}} \qquad \text{(II-242)}$$

Now that we have determined α and β in terms of h, τ_{max} will now be formulated. Since the contact is a conforming one, the equation for τ_{max} is

$$\tau_{max} \approx K \sqrt{1/2^2 + \mu^2} \ \frac{P}{A} \qquad \text{(II-243)}$$

where A is given by Eq. (II-222).

Using Eqs. (II-224) and (II-233), we get for A

$$A \approx 2\pi r \sqrt{R/R+r} \quad h \tag{II-244}$$

Therefore

$$\tau_{max} \approx \frac{K \sqrt{(R+r)(1/2^2 + \mu^2)} \quad P}{2\pi r R^{1/2} h} \tag{II-245}$$

$(\tau_{max} W)$ is obtained by combining Eqs. (II-242) and (II-245) and this results in

$$\tau_{max} W \approx \frac{P K \sqrt{1/2^2 + \mu^2}}{2^{1/2} \pi r^{1/2} h^{1/2}} \tag{II-246}$$

The differential for Q and $(\tau_{max} W)$ involved in Eq. (II-221) will now be determined by the use of Eqs. (II-226) and (II-246)

$$dQ = \frac{3}{\sqrt{2}} r^{1/2} h^{1/2} dh \tag{II-245}$$

$$d(\tau_{max} W) = - \frac{P K \sqrt{1/2^2 + \mu^2}}{2^{3/2} \pi r^{1/2} h^{3/2}} dh \tag{II-248}$$

The remaining differential to be formed is that for dN. In the discussion concerning the design procedure for non-zero wear, we see that the relationship between the differentials of the number of passes on the sphere and the number of revolutions is

$$dN = n_s dL \tag{II-249}$$

where

$$n_s = \frac{2\pi R}{W} \tag{II-250}$$

Using Eq. (II-242), for W, we get

$$n_s = \frac{\pi 2^{1/2} R^{1/2} (R+r)^{1/2}}{r^{1/2} h^{1/2}} \tag{II-251}$$

Therefore,

$$dN = \frac{\pi \, 2^{1/2} \, R^{1/2} \, (R+r)^{1/2}}{r^{1/2} \, h^{1/2}} \, dL \qquad (II\text{-}252)$$

It is now possible to substitute for all the quantities in Eq. (II-221) expressions involving only the parameters of the mechanism considered. If Eqs. (II-226), (II-246), (II-247), (II-248) and (II-252) are substituted into Eq. (II-221), we have, upon combination,

$$\frac{15}{4} \, (2r)^{1/2} \, h^{13/4} dh = \frac{C'' \, P^{9/2} \, K^{9/2} \, (1/2^2 + \mu^2)^{9/4} \, (R+r)^{1/2}}{2^{7/4} \, \pi^{7/2} \, r^{11/4}} \, dL \qquad (II\text{-}253)$$

This is the differential equation which describes the wear on the sphere. As can be seen, it can readily be integrated, which is the second step, to give

$$h^{17/4} = \frac{17C'' \, P^{9/2} \, K^{9/2} \, (1/2^2 + \mu^2)^{9/4} \, (R+r)^{1/2} \, L}{15 \cdot 2^{9/4} \, \pi^{7/2} \, r^{13/4}} + C_2 \qquad (II\text{-}254)$$

or it can be written as

$$h^{17/4} = C_1 L + C_2 \qquad (II\text{-}255)$$

where

$$C_1 = \frac{17C'' \, P^{9/2} \, K^{9/2} \, (1/2^2 + \mu^2)^{9/4} \, (R+r)^{1/2}}{15 \cdot 2^{9/4} \, \pi^{7/2} \, r^{13/4}} \qquad (II\text{-}256)$$

Before Eq. (II-255) can be used to determine whether the design of the mechanism is satisfactory from a wear standpoint, the dependency of the C's on various parameters must be determined. This is the third step. Since in the problem considered, there will be a certain lifetime in which there will be zero wear on the sphere, and in addition, since it can be inferred from Eq. (II-220) that this lifetime exceeds 2000 passes, the analytical determination of the C's will be used, although the alternate procedure of experimentally determining the C's may also be used. In the analytical procedure, C_2 is taken to equal zero and C_1 is determined from the surface finish of the material and the number of operations, L_1, for which there will be zero wear on the sphere. For the earlier discussion of the design application of non-zero wear model (Eq. (II-206)):

$$L_1 = \frac{2 \times 10^3}{n_s} \left(\frac{\gamma_R \, \tau_{ys}}{\tau_{max}} \right)^9 \qquad (II\text{-}257)$$

Even though this manner of determining the C's allows one to explicitly give the dependencies of C_1 on the radii, loads, yield points in shear, etc., the problem will not be formulated in that manner. Rather the value of C_1 for the given values of the parameters will be determined, since the aim of this analysis is to determine whether the present set of parameters is satisfactory. If the present set is not satisfactory, these dependencies will be introduced explicitly so that suitable values of the parameters which can be changed can be determined.

γ_R, τ_{ys} and τ_{max} have already been determined for the proposed conditions in the zero wear analysis portion of the problem. n_s was not determined specifically. This can be obtained by dividing N_s by the number of operations, 10^6.

$$n_s = \frac{4.98 \times 10^9}{10^6} \tag{II-258}$$

$$n_s = 4.98 \times 10^3 \tag{II-259}$$

Therefore

$$L_1 = \frac{2 \times 10^3}{4.98 \times 10^3} \left(\frac{0.54 \times 40 \times 10^3}{1.00 \times 10^4} \right)^9 \tag{II-260}$$

$$L_1 = 4.02 \times 10^2 \tag{II-261}$$

This means that in 4.02×10^2 operations, the wear scar has achieved a maximum depth of approximately one-half the peak-to-peak value of the surface finish. If we assume the peak-to-peak finish to be 30 μ inches, h_1, being the wear after L_1 operations, is 15 μ inches. C_1 is now determined for the particular conditions given by the means of Eq. (II-255), using for L and h, the values of L_1 and h_1.

$$C_1 = \frac{h_1^{17/4}}{L_1} \tag{II-262}$$

$$C_1 = \frac{(15 \times 10^{-6})^{17/4} \ in^{17/4}}{4.02 \times 10^2} \tag{II-263}$$

$$C_1 = 2.36 \times 10^{-23} \ in^{17/4} \tag{II-264}$$

Eq. (II-255) becomes for this mechanism with the present specifications of load, geometry, etc.,

$$h^{17/4} = 2.36 \times 10^{-23} \ L \ (in.^{17/4}) \tag{II-265}$$

For h = 0.005 inches, the number of operations required to achieve that amount of wear is determined by

$$L = \frac{(.005)^{17/4}}{2.36 \times 10^{-23}} \tag{II-266}$$

$$L = 6.45 \times 10^{12} \tag{II-267}$$

This means that this mechanism would be expected to operate for 6×10^{12} revolutions before the maximum depth of the groove in the sphere would reach a value of 0.005 inches. Since only 10^6 revolutions are required, the analysis indicates that the mechanism as specified will be satisfactory, and it will not be necessary to investigate changes in some of the parameters.

III TABLES

In this section, tables and figures which are pertinent to the design procedure discussed earlier are presented. Before these tables are discussed in detail, it is well to point out that the tables and figures should, in general, not be viewed as engineering standards, but instead should be viewed as tables of values which have been found to be suitable for application of the wear models in most design problems.

A listing of the contents of this section is as follows:

Figure III-1 is simply a summary of the expressions given in Section 2 for determining values of τ_{max} for several types of geometries. Table III-1 is associated with Figure III-1 and is concerned with the numerical relationship between m, n and $\cos\theta$ occurring in the formulation for the general Hertzian case. It lists the values of m and n for values of $\cos\theta$, ranging from 0.9999 to 0 with increments ranging from 0.0001 to 0.0020, depending upon the particular interval of $\cos\theta$.

Figure III-2 gives a simple nomograph for determining the ninth root of $2 \times 10^3/N$. The root is obtained in the following manner: Write N in the form $u.vw \times 10^y$. For example, if N = 12 million, write it as 1.2×10^7. Then select the appropriate curve by determining which curve includes in its range the value of N considered (see upper right-hand corner of Figure III-2). In this case, it would be curve 5. Enter the graph on the axis at the point $u.vw$ — in this case at 1.20. Then proceed from this point along a

line parallel to the $\left(\dfrac{2 \times 10^3}{N}\right)^{1/9}$ axis until the curve previously selected is encountered —
in the example considered, until curve 5 is encountered. Then draw a line parallel to the
N axis extending from the point of intersection to the $(2 \times 10^3/N)^{1/9}$ axis and read the value
of the root at the point where this line intersects the axis. In the case considered, it is 0.384.

Figure III-3 presents a graph relating τ_y to H_m. For the purposes of the design procedure, this curve may be used for all materials and has been found to be accurate for this purpose
to within 10%.

Table III-2 gives the values of Young's modulus and Poisson's ratio for a large number
of materials. For materials which are contained in Tables III-3 and III-4, the values of E and
ν used in determining γ_R are the ones listed in this table.

Table III-3 contains values of γ_R for various combinations of materials and lubricants.
This table is divided into three main categories based on materials. The first category is one
in which one member of the combination is 52100 steel. The second is one in which one
member is 302 stainless steel and the third is one in which brass is a member. In addition to
this, at the end there are a few combinations in which sintered materials and steels appear
but which do not come under the above categories. In each category, the data is further
subdivided on the basis of the other materials of the combination into the following sequence:
Combinations with stainless steels, steels, nickel alloys, copper alloys, aluminum alloys,
sintered brass, sintered bronze, sintered iron, sintered iron-copper, sintered steel, layered
materials, and plastics. In each of these categories, not all of these subdivisions appear.

A description of the materials which appear in Table III-3 is given in Table III-4.
These descriptions include chemical composition as well as the microhardness and yield point
in shear of the materials evaluated. The materials in Table III-4 are listed in the same
order as the subdivision in Table III-3.

In Table III-3, for each combination of materials, the values of γ_R and μ for several
lubricants are given. The manner of lubrication is specified as: dry, A, B, C or D. When
it is specified as dry, this means that the value of γ_R is appropriate when no lubricant is used.
When A, B, C or D is specified, this means that the value of γ_R is appropriate for conditions
where lubricant A, B, C or D, respectively, is used under conditions of boundary lubrication.
The description of the lubricants A, B, C and D is found in Table III-5.

It is well to emphasize here that unless a value of $\gamma_R = 0.2$ for a given combination
of materials and lubricant is assumed, Tables III-3, III-4 and III-5 must be used in conjunction
with one another to determine the appropriate value of γ_R, since γ_R is a function of the
composition of the lubricant and the materials of the combination for dry, clean conditions
or boundary lubrication conditions.

General Hertzian Case

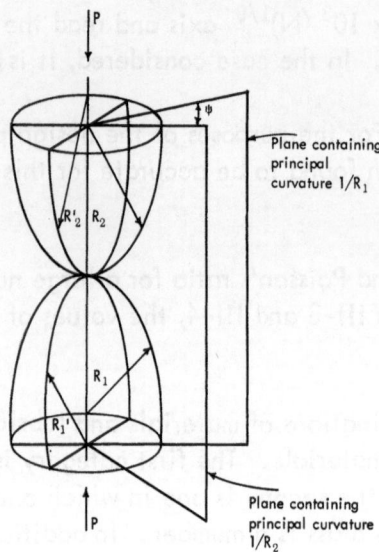

Plane containing principal curvature $1/R_1$

Plane containing principal curvature $1/R_2$

SYMBOLS

μ = Coefficient of friction

ν = Poisson's Ratio

E = Young's Modulus

K = Stress concentration factor

Parallel Cylinder

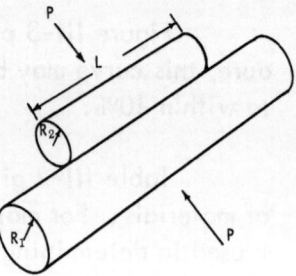

If sliding is parallel to a-axis, τ_{max} is given by largest of

$$\tau = q_0 \left(\frac{1-2\nu}{3}\right) \frac{b}{a}$$

$$\tau = q_0 \sqrt{\frac{1}{4}\left[(1-2\nu)\frac{a}{a+b}\right]^2 + \mu^2}$$

$$\tau = 0.31\, q_0$$

If sliding is parallel to b-axis, τ_{max} is given by largest of

$$\tau = q_0 \left(\frac{1-2\nu}{3}\right) \frac{b}{a}$$

$$\tau = q_0 \sqrt{\frac{1}{4}\left[(1-2\nu)\frac{b}{(a+b)}\right]^2 + \mu^2}$$

$$\tau = 0.31\, q_0$$

where

$$q_0 = \frac{3}{2} \frac{P}{\pi a b}$$

$$a = m \sqrt[3]{\frac{3\pi}{4} \frac{P(k_1+k_2)}{(B+A)}} \qquad b = n \sqrt[3]{\frac{3\pi}{4} \frac{P(k_1+k_2)}{(B+A)}}$$

m and n are found from $\cos\theta$ by means of Table III-1.

$$\cos\theta = (B-A)/(B+A)$$

$$B-A = \frac{1}{2}\left[\left(\frac{1}{R_1}-\frac{1}{R'_1}\right)^2 + \left(\frac{1}{R_2}-\frac{1}{R'_2}\right)^2 + 2\left(\frac{1}{R_1}-\frac{1}{R'_1}\right)\left(\frac{1}{R_2}-\frac{1}{R'_2}\right)\cos 2\psi\right]^{1/2}$$

$$B+A = \frac{1}{2}\left(\frac{1}{R_1}+\frac{1}{R'_1}+\frac{1}{R_2}+\frac{1}{R'_2}\right)$$

$$k = \frac{1-\nu^2}{\pi E}$$

If sliding is parallel to axis of cylinder

$$\tau_{max} = K q_0 \sqrt{\frac{1}{2}^2 + \mu^2}$$

If sliding is perpendicular to axis of cylinder

$$\tau_{max} = K \left(\frac{1+\mu}{2}\right) q_0$$

where

$$q_0 = \frac{2}{\pi} \frac{P}{Lb}$$

$$b = 2 \sqrt{\frac{P}{L} \frac{R_1 R_2 (k_1+k_2)}{R_1+R_2}}$$

$$k = \frac{1-\nu^2}{\pi E}$$

Conforming Geometries

$$\tau_{max} = K q_0 \sqrt{\frac{1}{2}^2 + \mu^2}$$

where

$$q_0 = P/A$$

A is the area of the contact projected onto a plane perpendicular to the line of action of P.

Fig. III-1 - Expression For τ_{max}

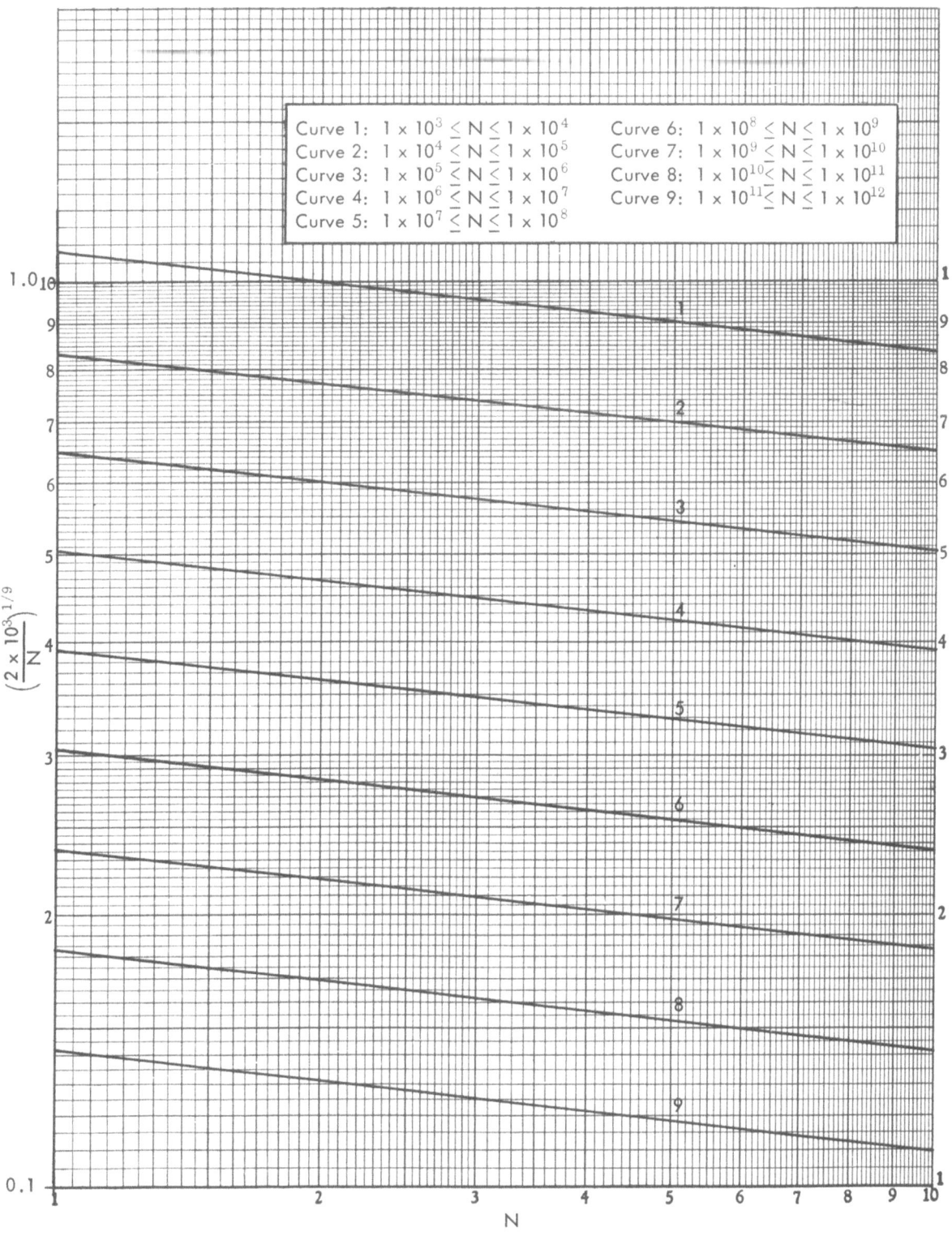

Fig. III-2 - Nomograph For Determining $\left(\dfrac{2 \times 10^3}{N}\right)^{1/9}$

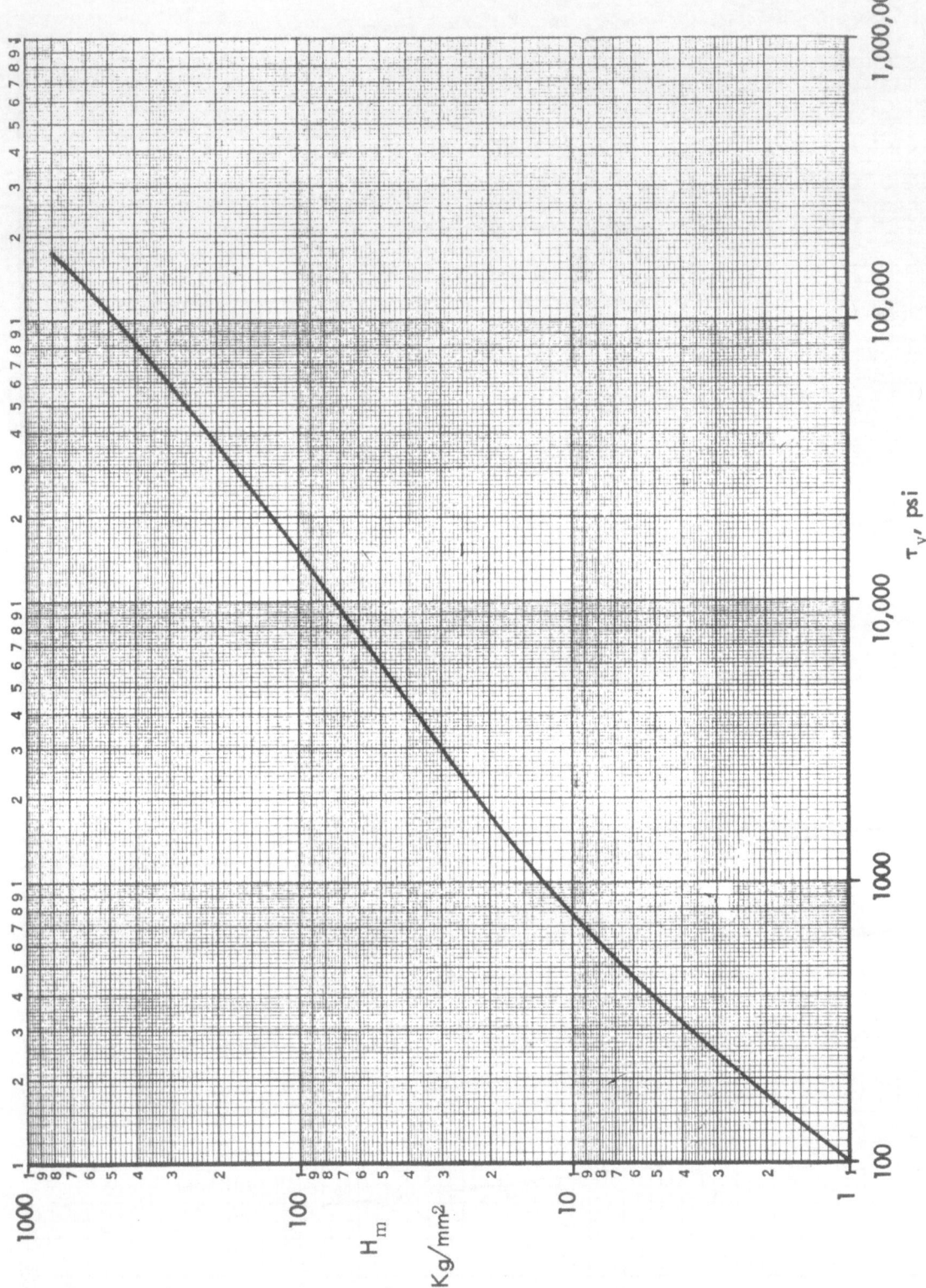

Fig. III-3 – Graph Relating H$_m$ to τ_y

TABLE III-1

Values of m and n in terms of cos θ

COS (θ)	m	n	COS (θ)	m	n
0.9999	39.93	0.1263	0.9945	9.764	0.2555
0.9998	31.69	0.1417	0.9944	9.694	0.2565
0.9997	27.69	0.1516	0.9943	9.648	0.2571
0.9996	25.15	0.1591	9.9942	9.581	0.2580
0.9995	23.35	0.1651	0.9941	9.516	0.2589
0.9994	21.98	0.1702	0.9940	9.453	0.2597
0.9993	20.87	0.1747	0.9939	9.391	0.2606
0.9992	19.97	0.1786	0.9938	9.331	0.2614
0.9991	19.20	0.1821	0.9937	9.272	0.2623
0.9990	18.54	0.1854	0.9936	9.234	0.2628
0.9989	17.96	0.1883	0.9935	9.177	0.2636
0.9988	17.44	0.1911	0.9934	9.123	0.2644
0.9987	16.98	0.1936	0.9933	9.069	0.2652
0.9986	16.38	0.1972	0.9932	9.017	0.2660
0.9985	15.85	0.2005	0.9931	8.966	0.2667
0.9984	15.53	0.2025	0.9930	8.916	0.2675
0.9983	15.24	0.2044	0.9929	8.867	0.2682
0.9982	14.97	0.2063	0.9928	8.819	0.2689
0.9981	14.59	0.2089	0.9927	8.772	0.2696
0.9980	14.25	0.2114	0.9926	8.726	0.2704
0.9979	14.04	0.2130	0.9925	8.682	0.2710
0.9978	13.85	0.2145	0.9924	8.638	0.2718
0.9977	13.57	0.2167	0.9923	8.594	0.2724
0.9976	13.31	0.2188	0.9922	8.552	0.2731
0.9975	13.15	0.2201	0.9921	8.511	0.2738
0.9974	13.00	0.2214	0.9920	8.470	0.2745
0.9973	12.78	0.2233	0.9919	8.430	0.2751
0.9972	12.58	0.2250	0.9918	8.391	0.2758
0.9971	12.45	0.2262	0.9917	8.352	0.2764
0.9970	12.33	0.2273	0.9916	8.315	0.2770
0.9969	12.15	0.2290	0.9915	8.278	0.2776
0.9968	11.99	0.2306	0.9914	8.242	0.2783
0.9967	11.88	0.2316	0.9913	8.206	0.2789
0.9966	11.73	0.2331	0.9912	8.171	0.2795
0.9965	11.58	0.2345	0.9911	8.136	0.2801
0.9964	11.49	0.2355	0.9910	8.091	0.2809
0.9963	11.35	0.2369	0.9909	8.058	0.2814
0.9962	11.23	0.2382	0.9908	8.025	0.2820
0.9961	11.15	0.2391	0.9907	7.993	0.2826
0.9960	11.03	0.2404	0.9906	7.961	0.2831
0.9959	10.92	0.2416	0.9905	7.930	0.2837
0.9958	10.84	0.2425	0.9904	7.899	0.2843
0.9957	10.73	0.2437	0.9903	7.869	0.2848
0.9956	10.63	0.2448	0.9902	7.839	0.2853
0.9955	10.53	0.2460	0.9901	7.800	0.2861
0.9954	10.44	0.2471	0.9900	7.772	0.2866
0.9953	10.38	0.2478	0.9899	7.743	0.2871
0.9952	10.29	0.2488	0.9898	7.716	0.2877
0.9951	10.20	0.2499	0.9897	7.688	0.2882
0.9950	10.15	0.2506	0.9896	7.661	0.2887
0.9949	10.06	0.2516	0.9895	7.634	0.2892
0.9948	9.987	0.2527	0.9894	7.599	0.2898
0.9947	9.910	0.2536	0.9893	7.573	0.2904
0.9946	9.836	0.2546	0.9892	7.548	0.2908

TABLE III-1 (Cont.)

COS (θ)	m	n	COS (θ)	m	n
0.9891	7.523	0.2914	0.9837	6.430	0.3154
0.9890	7.498	0.2918	0.9836	6.418	0.3157
0.9889	7.465	0.2924	0.9835	6.404	0.3160
0.9888	7.441	0.2930	0.9834	6.387	0.3165
0.9887	7.417	0.2934	0.9833	6.370	0.3169
0.9886	7.386	0.2940	0.9832	6.357	0.3172
0.9885	7.363	0.2945	0.9831	6.344	0.3175
0.9884	7.340	0.2949	0.9830	6.328	0.3180
0.9883	7.318	0.2954	0.9829	6.311	0.3184
0.9882	7.295	0.2958	0.9828	6.299	0.3187
0.9881	7.266	0.2964	0.9827	6.287	0.3190
0.9880	7.245	0.2969	0.9826	6.271	0.3194
0.9879	7.223	0.2973	0.9825	6.255	0.3198
0.9878	7.195	0.2979	0.9824	6.243	0.3201
0.9877	7.174	0.2984	0.9823	6.231	0.3205
0.9876	7.154	0.2988	0.9822	6.215	0.3209
0.9875	7.134	0.2993	0.9821	6.200	0.3213
0.9874	7.113	0.2997	0.9820	6.189	0.3216
0.9873	7.087	0.3002	0.9819	6.177	0.3219
0.9872	7.067	0.3007	0.9818	6.162	0.3223
0.9871	7.048	0.3011	0.9817	6.147	0.3227
0.9870	7.022	0.3016	0.9816	6.132	0.3230
0.9869	7.004	0.3021	0.9815	6.122	0.3233
0.9868	6.985	0.3024	0.9814	6.110	0.3236
0.9867	6.960	0.3030	0.9813	6.096	0.3240
0.9866	6.942	0.3034	0.9812	6.082	0.3244
0.9865	6.924	0.3038	0.9811	6.067	0.3248
0.9864	6.900	0.3043	0.9810	6.057	0.3251
0.9863	6.883	0.3047	0.9809	6.047	0.3254
0.9862	6.865	0.3051	0.9808	6.033	0.3257
0.9861	6.842	0.3056	0.9807	6.019	0.3261
0.9860	6.826	0.3060	0.9806	6.009	0.3264
0.9859	6.808	0.3064	0.9805	5.999	0.3267
0.9858	6.786	0.3069	0.9804	5.986	0.3270
0.9857	6.770	0.3073	0.9803	5.972	0.3274
0.9856	6.753	0.3076	0.9802	5.959	0.3278
0.9855	6.732	0.3081	0.9801	5.946	0.3281
0.9854	6.711	0.3086	0.9800	5.937	0.3284
0.9853	6.695	0.3090	0.9799	5.927	0.3287
0.9852	6.680	0.3093	0.9798	5.914	0.3290
0.9851	6.659	0.3098	0.9797	5.902	0.3294
0.9850	6.644	0.3102	0.9796	5.889	0.3297
0.9849	6.629	0.3106	0.9795	5.880	0.3300
0.9848	6.609	0.3110	0.9794	5.871	0.3303
0.9847	6.594	0.3114	0.9793	5.858	0.3306
0.9846	6.579	0.3117	0.9792	5.846	0.3310
0.9845	6.560	0.3122	0.9791	5.834	0.3313
0.9844	6.541	0.3127	0.9790	5.822	0.3317
0.9843	6.527	0.3130	0.9789	5.813	0.3319
0.9842	6.512	0.3133	0.9788	5.805	0.3322
0.9841	6.494	0.3138	0.9787	5.793	0.3225
0.9840	6.480	0.3141	0.9786	5.781	0.3329
0.9839	6.466	0.3145	0.9785	5.770	0.3332
0.9838	6.448	0.3149	0.9784	5.758	0.3335

TABLE III-1 (Cont.)

COS (θ)	m	n	COS (θ)	m	n
0.9783	5.750	0.3338	0.9729	5.266	0.3491
0.9782	5.741	0.3340	0.9728	5.258	0.3493
0.9781	5.730	0.3344	0.9727	5.250	0.3496
0.9780	5.719	0.3347	0.9726	5.242	0.3498
0.9779	5.708	0.3350	0.9725	5.236	0.3501
0.9778	5.697	0.3353	0.9724	5.230	0.3503
0.9777	5.686	0.3357	0.9723	5.223	0.3505
0.9776	5.678	0.3359	0.9722	5.215	0.3508
0.9775	5.670	0.3361	0.9721	5.207	0.3510
0.9774	5.659	0.3365	0.9720	5.200	0.3513
0.9773	5.649	0.3368	0.9719	5.192	0.3516
0.9772	5.638	0.3371	0.9718	5.185	0.3518
0.9771	5.628	0.3374	0.9717	5.177	0.3521
0.9770	5.617	0.3377	0.9716	5.170	0.3523
0.9769	5.610	0.3380	0.9715	5.163	0.3526
0.9768	5.602	0.3382	0.9714	5.155	0.3528
0.9767	5.592	0.3385	0.9713	5.148	0.3531
0.9766	5.582	0.3388	0.9712	5.141	0.3533
0.9765	5.572	0.3391	0.9711	5.134	0.3536
0.9764	5.562	0.3394	0.9710	5.126	0.3538
0.9763	5.552	0.3398	0.9709	5.119	0.3541
0.9762	5.542	0.3401	0.9708	5.112	0.3543
0.9761	5.535	0.3403	0.9707	5.105	0.3546
0.9760	5.528	0.3405	0.9706	5.098	0.3548
0.9759	5.518	0.3408	0.9705	5.091	0.3551
0.9758	5.508	0.3411	0.9704	5.084	0.3553
0.9757	5.499	0.3414	0.9703	5.077	0.3556
0.9756	5.490	0.3417	0.9702	5.071	0.3558
0.9755	5.480	0.3420	0.9701	5.064	0.3561
0.9754	5.471	0.3423	0.9700	5.057	0.3563
0.9753	5.462	0.3426	0.9699	5.050	0.3566
0.9752	5.455	0.3428	0.9698	5.044	0.3568
0.9751	5.448	0.3430	0.9697	5.037	0.3570
0.9750	5.439	0.3433	0.9696	5.030	0.3573
0.9749	5.430	0.3436	0.9695	5.024	0.3575
0.9748	5.421	0.3439	0.9694	5.017	0.3578
0.9747	5.412	0.3442	0.9693	5.011	0.3580
0.9746	5.403	0.3445	0.9692	5.004	0.3582
0.9745	5.395	0.3448	0.9691	4.998	0.3585
0.9744	5.386	0.3451	0.9690	4.991	0.3587
0.9743	5.377	0.3453	0.9689	4.985	0.3589
0.9742	5.369	0.3456	0.9688	4.979	0.3592
0.9741	5.360	0.3459	0.9687	4.972	0.3594
0.9740	5.354	0.3461	0.9686	4.966	0.3596
0.9739	5.347	0.3464	0.9685	4.960	0.3599
0.9738	5.339	0.3466	0.9684	4.953	0.3601
0.9737	5.331	0.3469	0.9683	4.947	0.3603
0.9736	5.322	0.3471	0.9682	4.941	0.3606
0.9735	5.314	0.3474	0.9681	4.935	0.3608
0.9734	5.306	0.3477	0.9680	4.929	0.3610
0.9733	5.298	0.3480	0.9679	4.923	0.3612
0.9732	5.290	0.3482	0.9678	4.917	0.3615
0.9731	5.282	0.3485	0.9677	4.911	0.3617
0.9730	5.274	0.3488	0.9676	4.905	0.3619

TABLE III-1 (Cont.)

COS (θ)	m	n	COS (θ)	m	n
0.9675	4.899	0.3621	0.9621	4.607	0.3738
0.9674	4.893	0.3624	0.9620	4.602	0.3740
0.9673	4.887	0.3626	0.9619	4.597	0.3742
0.9672	4.881	0.3628	0.9618	4.593	0.3744
0.9671	4.875	0.3630	0.9617	4.588	0.3746
0.9670	4.870	0.3633	0.9616	4.583	0.3748
0.9669	4.864	0.3635	0.9615	4.577	0.3750
0.9668	4.858	0.3637	0.9614	4.572	0.3752
0.9667	4.852	0.3639	0.9613	4.568	0.3754
0.9666	4.847	0.3641	0.9612	4.564	0.3756
0.9665	4.841	0.3644	0.9611	4.559	0.3758
0.9664	4.835	0.3646	0.9610	4.555	0.3760
0.9663	4.830	0.3648	0.9609	4.550	0.3762
0.9662	4.823	0.3651	0.9608	4.545	0.3764
0.9661	4.816	0.3653	0.9607	4.539	0.3766
0.9660	4.810	0.3655	0.9606	4.534	0.3769
0.9659	4.805	0.3658	0.9604	4.528	0.3772
0.9658	4.799	0.3660	0.9603	4.521	0.3774
0.9657	4.794	0.3662	0.9601	4.515	0.3777
0.9656	4.789	0.3664	0.9600	4.508	0.3780
0.9655	4.783	0.3666	0.9599	4.502	0.3783
0.9654	4.778	0.3668	0.9597	4.496	0.3785
0.9653	4.773	0.3670	0.9596	4.489	0.3788
0.9652	4.767	0.3672	0.9594	4.483	0.3791
0.9651	4.762	0.3675	0.9593	4.477	0.3794
0.9650	4.757	0.3677	0.9592	4.471	0.3796
0.9649	4.751	0.3679	0.9590	4.465	0.3799
0.9648	4.746	0.3681	0.9589	4.459	0.3802
0.9647	4.741	0.3683	0.9587	4.453	0.3804
0.9646	4.735	0.3686	0.9586	4.447	0.3807
0.9645	4.728	0.3688	0.9585	4.441	0.3810
0.9644	4.723	0.3690	0.9583	4.435	0.3812
0.9643	4.718	0.3692	0.9582	4.429	0.3815
0.9642	4.713	0.3694	0.9581	4.423	0.3817
0.9641	4.708	0.3696	0.9579	4.417	0.3820
0.9640	4.703	0.3698	0.9578	4.411	0.3823
0.9639	4.698	0.3700	0.9576	4.405	0.3825
0.9638	4.693	0.3702	0.9575	4.400	0.3828
0.9637	4.688	0.3704	0.9574	4.394	0.3830
0.9636	4.683	0.3706	0.9572	4.388	0.3833
0.9635	4.678	0.3708	0.9571	4.382	0.3836
0.9634	4.672	0.3711	0.9570	4.377	0.3838
0.9633	4.666	0.3713	0.9568	4.371	0.3841
0.9632	4.661	0.3715	0.9567	4.366	0.3843
0.9631	4.656	0.3717	0.9565	4.360	0.3846
0.9630	4.651	0.3719	0.9564	4.355	0.3848
0.9629	4.646	0.3721	0.9563	4.349	0.3851
0.9628	4.642	0.3723	0.9561	4.344	0.3853
0.9627	4.637	0.3725	0.9560	4.338	0.3856
0.9626	4.632	0.3727	0.9559	4.333	0.3858
0.9625	4.627	0.3729	0.9557	4.327	0.3861
0.9624	4.622	0.3732	0.9556	4.322	0.3863
0.9623	4.616	0.3734	0.9555	4.317	0.3866
0.9622	4.611	0.3736	0.9553	4.311	0.3868

TABLE III-1 (Cont.)

COS (θ)	m	n	COS (θ)	m	n
0.9552	4.306	0.3871	0.9480	4.055	0.3993
0.9550	4.301	0.3873	0.9479	4.050	0.3995
0.9549	4.296	0.3876	0.9478	4.046	0.3997
0.9548	4.290	0.3878	0.9476	4.042	0.4000
0.9546	4.285	0.3880	0.9475	4.038	0.4002
0.9545	4.280	0.3883	0.9474	4.034	0.4004
0.9544	4.275	0.3885	0.9472	4.030	0.4006
0.9542	4.270	0.3888	0.9471	4.026	0.4008
0.9541	4.265	0.3890	0.9470	4.022	0.4010
0.9540	4.260	0.3892	0.9468	4.018	0.4012
0.9538	4.255	0.3895	0.9467	4.014	0.4014
0.9537	4.250	0.3897	0.9463	4.001	0.4021
0.9536	4.245	0.3899	0.9459	3.988	0.4028
0.9534	4.240	0.3902	0.9454	3.975	0.4034
0.9533	4.235	0.3904	0.9450	3.963	0.4041
0.9532	4.230	0.3907	0.9446	3.950	0.4048
0.9530	4.225	0.3909	0.9442	3.938	0.4054
0.9529	4.220	0.3911	0.9437	3.926	0.4061
0.9528	4.215	0.3914	0.9433	3.914	0.4067
0.9526	4.210	0.3916	0.9429	3.902	0.4074
0.9525	4.206	0.3918	0.9425	3.890	0.4080
0.9524	4.201	0.3920	0.9420	3.879	0.4087
0.9522	4.196	0.3923	0.9416	3.867	0.4093
0.9521	4.191	0.3925	0.9412	3.856	0.4099
0.9520	4.187	0.3927	0.9408	3.845	0.4105
0.9518	4.182	0.3930	0.9404	3.834	0.4111
0.9517	4.177	0.3932	0.9400	3.823	0.4118
0.9516	4.173	0.3934	0.9395	3.812	0.4124
0.9514	4.168	0.3936	0.9391	3.802	0.4130
0.9513	4.163	0.3939	0.9387	3.791	0.4136
0.9512	4.159	0.3941	0.9383	3.781	0.4142
0.9510	4.154	0.3943	0.9379	3.770	0.4147
0.9509	4.150	0.3945	0.9375	3.760	0.4153
0.9508	4.145	0.3948	0.9371	3.750	0.4159
0.9506	4.141	0.3950	0.9367	3.740	0.4165
0.9505	4.136	0.3952	0.9363	3.730	0.4171
0.9504	4.132	0.3954	0.9359	3.721	0.4176
0.9502	4.127	0.3957	0.9355	3.711	0.4182
0.9501	4.123	0.3959	0.9350	3.702	0.4188
0.9500	4.118	0.3961	0.9346	3.692	0.4193
0.9498	4.114	0.3963	0.9342	3.683	0.4199
0.9497	4.110	0.3965	0.9338	3.674	0.4205
0.9496	4.105	0.3968	0.9334	3.664	0.4210
0.9494	4.101	0.3970	0.9330	3.655	0.4216
0.9493	4.097	0.3972	0.9326	3.646	0.4221
0.9492	4.092	0.3974	0.9322	3.637	0.4226
0.9491	4.088	0.3976	0.9318	3.629	0.4232
0.9489	4.084	0.3978	0.9314	3.620	0.4237
0.9488	4.080	0.3980	0.9310	3.611	0.4242
0.9487	4.075	0.3983	0.9306	3.603	0.4248
0.9485	4.071	0.3985	0.9303	3.594	0.4253
0.9484	4.067	0.3987	0.9299	3.586	0.4258
0.9483	4.063	0.3989	0.9295	3.578	0.4263
0.9481	4.059	0.3991	0.9291	3.570	0.4269

TABLE III-1 (Cont.)

COS (θ)	m	n	COS (θ)	m	n
0.9287	3.561	0.4274	0.9081	3.204	0.4520
0.9283	3.553	0.4279	0.9077	3.199	0.4524
0.9279	3.545	0.4284	0.9074	3.194	0.4528
0.9275	3.537	0.4289	0.9070	3.189	0.4532
0.9271	3.530	0.4294	0.9066	3.184	0.4536
0.9267	3.522	0.4299	0.9063	3.179	0.4540
0.9263	3.514	0.4304	0.9059	3.174	0.4544
0.9259	3.506	0.4309	0.9056	3.168	0.4548
0.9256	3.499	0.4314	0.9052	3.163	0.4551
0.9252	3.491	0.4319	0.9049	3.159	0.4555
0.9248	3.484	0.4324	0.9045	3.154	0.4559
0.9244	3.477	0.4328	0.9042	3.149	0.4563
0.9240	3.469	0.4333	0.9038	3.144	0.4567
0.9236	3.462	0.4338	0.9035	3.139	0.4570
0.9233	3.455	0.4343	0.9031	3.134	0.4574
0.9229	3.448	0.4347	0.9028	3.129	0.4578
0.9225	3.441	0.4352	0.9024	3.125	0.4582
0.9221	3.434	0.4357	0.9021	3.120	0.4585
0.9217	3.427	0.4362	0.9017	3.115	0.4589
0.9213	3.420	0.4366	0.9014	3.111	0.4593
0.9210	3.413	0.4371	0.9010	3.106	0.4597
0.9206	3.406	0.4375	0.9007	3.101	0.4600
0.9202	3.400	0.4380	0.9003	3.097	0.4604
0.9198	3.393	0.4385	0.9000	3.092	0.4607
0.9195	3.386	0.4389	0.8996	3.088	0.4611
0.9191	3.380	0.4394	0.8993	3.083	0.4615
0.9187	3.373	0.4398	0.8989	3.079	0.4618
0.9183	3.367	0.4403	0.8986	3.074	0.4622
0.9180	3.360	0.4407	0.8982	3.070	0.4625
0.9176	3.354	0.4412	0.8979	3.066	0.4629
0.9172	3.348	0.4416	0.8975	3.061	0.4633
0.9168	3.341	0.4420	0.8972	3.057	0.4636
0.9165	3.335	0.4425	0.8969	3.053	0.4640
0.9161	3.329	0.4429	0.8965	3.048	0.4643
0.9157	3.323	0.4433	0.8962	3.044	0.4647
0.9154	3.317	0.4438	0.8958	3.040	0.4650
0.9150	3.311	0.4442	0.8955	3.036	0.4654
0.9146	3.305	0.4446	0.8951	3.032	0.4657
0.9143	3.299	0.4451	0.8948	3.027	0.4661
0.9139	3.293	0.4455	0.8945	3.023	0.4664
0.9135	3.287	0.4459	0.8941	3.019	0.4667
0.9132	3.281	0.4463	0.8938	3.015	0.4671
0.9128	3.276	0.4467	0.8934	3.011	0.4674
0.9124	3.270	0.4472	0.8931	3.007	0.4678
0.9121	3.264	0.4476	0.8928	3.003	0.4681
0.9117	3.259	0.4480	0.8924	2.999	0.4684
0.9113	3.253	0.4484	0.8921	2.995	0.4688
0.9110	3.247	0.4488	0.8918	2.991	0.4691
0.9106	3.242	0.4492	0.8914	2.987	0.4695
0.9102	3.236	0.4496	0.8911	2.983	0.4698
0.9099	3.231	0.4500	0.8907	2.979	0.4701
0.9095	3.226	0.4504	0.8904	2.975	0.4704
0.9092	3.220	0.4508	0.8901	2.972	0.4708
0.9088	3.215	0.4512	0.8897	2.968	0.4711
0.9084	3.210	0.4516	0.8894	2.964	0.4714

TABLE III-1 (Cont.)

COS (θ)	m	n	COS (θ)	m	n
0.8891	2.960	0.4718	0.8713	2.778	0.4884
0.8887	2.956	0.4721	0.8710	2.775	0.4887
0.8884	2.953	0.4724	0.8707	2.773	0.4889
0.8881	2.949	0.4727	0.8704	2.770	0.4892
0.8877	2.945	0.4731	0.8701	2.767	0.4895
0.8874	2.941	0.4734	0.8698	2.764	0.4898
0.8871	2.938	0.4737	0.8694	2.761	0.4901
0.8868	2.934	0.4740	0.8691	2.758	0.4903
0.8864	2.931	0.4743	0.8688	2.756	0.4906
0.8861	2.927	0.4747	0.8685	2.753	0.4909
0.8858	2.923	0.4750	0.8682	2.750	0.4912
0.8854	2.920	0.4753	0.8679	2.747	0.4914
0.8851	2.916	0.4756	0.8676	2.744	0.4917
0.8848	2.913	0.4759	0.8673	2.742	0.4920
0.8844	2.909	0.4762	0.8670	2.739	0.4922
0.8841	2.906	0.4766	0.8667	2.736	0.4925
0.8838	2.902	0.4769	0.8663	2.733	0.4928
0.8835	2.899	0.4772	0.8660	2.731	0.4930
0.8831	2.895	0.4775	0.8657	2.728	0.4933
0.8828	2.892	0.4778	0.8654	2.725	0.4936
0.8825	2.888	0.4781	0.8651	2.723	0.4939
0.8822	2.885	0.4784	0.8648	2.720	0.4941
0.8818	2.882	0.4787	0.8645	2.717	0.4944
0.8815	2.878	0.4790	0.8642	2.715	0.4947
0.8812	2.875	0.4793	0.8639	2.712	0.4949
0.8809	2.871	0.4796	0.8636	2.710	0.4952
0.8805	2.868	0.4799	0.8633	2.707	0.4954
0.8802	2.865	0.4802	0.8630	2.704	0.4957
0.8799	2.862	0.4805	0.8627	2.702	0.4960
0.8796	2.858	0.4808	0.8624	2.699	0.4962
0.8792	2.855	0.4811	0.8621	2.697	0.4965
0.8789	2.852	0.4814	0.8618	2.694	0.4968
0.8786	2.849	0.4817	0.8615	2.691	0.4970
0.8783	2.845	0.4820	0.8612	2.689	0.4973
0.8780	2.842	0.4823	0.8609	2.686	0.4975
0.8776	2.839	0.4826	0.8605	2.684	0.4978
0.8773	2.836	0.4829	0.8602	2.681	0.4980
0.8770	2.833	0.4832	0.8599	2.679	0.4983
0.8767	2.830	0.4835	0.8596	2.676	0.4986
0.8764	2.826	0.4838	0.8593	2.674	0.4988
0.8760	2.823	0.4841	0.8590	2.671	0.4991
0.8757	2.820	0.4844	0.8587	2.669	0.4993
0.8754	2.817	0.4847	0.8584	2.667	0.4996
0.8751	2.814	0.4850	0.8581	2.664	0.4998
0.8748	2.811	0.4853	0.8578	2.662	0.5001
0.8745	2.808	0.4855	0.8575	2.659	0.5003
0.8741	2.805	0.4858	0.8572	2.657	0.5006
0.8738	2.802	0.4861	0.8569	2.654	0.5008
0.8735	2.799	0.4864	0.8566	2.652	0.5011
0.8732	2.796	0.4867	0.8563	2.650	0.5013
0.8729	2.793	0.4870	0.8561	2.647	0.5016
0.8726	2.790	0.4873	0.8558	2.645	0.5018
0.8723	2.787	0.4875	0.8555	2.643	0.5021
0.8719	2.784	0.4878	0.8552	2.640	0.5023
0.8716	2.781	0.4881	0.8549	2.638	0.5026

TABLE   III-1 (Cont.)

COS (θ)	m	n	COS (θ)	m	n
0.8546	2.636	0.5028	0.8387	2.519	0.5157
0.8543	2.633	0.5031	0.8384	2.517	0.5159
0.8540	2.631	0.5033	0.8381	2.516	0.5161
0.8537	2.629	0.5036	0.8378	2.514	0.5165
0.8534	2.626	0.5038	0.8375	2.512	0.5166
0.8531	2.624	0.5041	0.8373	2.510	0.5168
0.8528	2.622	0.5043	0.8370	2.508	0.5170
0.8525	2.620	0.5046	0.8367	2.506	0.5172
0.8522	2.617	0.5048	0.8364	2.504	0.5175
0.8519	2.615	0.5050	0.8361	2.502	0.5177
0.8516	2.613	0.5053	0.8359	2.500	0.5179
0.8513	2.611	0.5055	0.8356	2.499	0.5181
0.8510	2.608	0.5058	0.8353	2.497	0.5183
0.8507	2.606	0.5060	0.8350	2.495	0.5185
0.8504	2.604	0.5062	0.8347	2.493	0.5188
0.8502	2.602	0.5065	0.8345	2.491	0.5190
0.8499	2.599	0.5067	0.8342	2.489	0.5192
0.8496	2.597	0.5070	0.8339	2.488	0.5194
0.8493	2.595	0.5072	0.8336	2.486	0.5196
0.8490	2.593	0.5074	0.8334	2.484	0.5198
0.8487	2.591	0.5077	0.8331	2.482	0.5201
0.8484	2.589	0.5079	0.8328	2.480	0.5203
0.8481	2.586	0.5081	0.8325	2.478	0.5205
0.8478	2.584	0.5084	0.8322	2.477	0.5207
0.8475	2.582	0.5086	0.8320	2.475	0.5209
0.8472	2.580	0.5089	0.8317	2.473	0.5211
0.8470	2.578	0.5091	0.8314	2.471	0.5213
0.8467	2.576	0.5093	0.8311	2.470	0.5215
0.8464	2.574	0.5096	0.8309	2.468	0.5218
0.8461	2.572	0.5098	0.8306	2.466	0.5220
0.8458	2.569	0.5100	0.8303	2.464	0.5222
0.8455	2.567	0.5103	0.8300	2.463	0.5224
0.8452	2.565	0.5105	0.8298	2.461	0.5226
0.8449	2.563	0.5107	0.8295	2.459	0.5228
0.8446	2.561	0.5109	0.8292	2.457	0.5230
0.8444	2.559	0.5112	0.8289	2.456	0.5232
0.8441	2.557	0.5114	0.8287	2.454	0.5234
0.8438	2.555	0.5116	0.8284	2.452	0.5236
0.8435	2.553	0.5119	0.8281	2.450	0.5238
0.8432	2.551	0.5121	0.8279	2.449	0.5241
0.8429	2.549	0.5123	0.8276	2.447	0.5243
0.8426	2.547	0.5126	0.8273	2.445	0.5245
0.8424	0.545	0.5128	0.8270	2.444	0.5247
0.8421	2.543	0.5130	0.8268	2.442	0.5249
0.8418	2.541	0.5132	0.8265	2.440	0.5251
0.8415	2.539	0.5135	0.8262	2.439	0.5253
0.8412	2.537	0.5137	0.8259	2.437	0.5255
0.8409	2.535	0.5139	0.8257	2.435	0.5257
0.8407	2.533	0.5141	0.8254	2.434	0.5259
0.8404	2.531	0.5144	0.8251	2.432	0.5261
0.8401	2.529	0.5146	0.8249	2.430	0.5263
0.8398	2.527	0.5148	0.8246	2.429	0.5265
0.8395	2.525	0.5150	0.8243	2.427	0.5267
0.8392	2.523	0.5153	0.8241	2.425	0.5269
0.8390	2.521	0.5155	0.8238	2.424	0.5271

TABLE III-1 (Cont.)

COS (θ)	m	n	COS (θ)	m	n
0.8235	2.422	0.5273	0.7493	2.070	0.5780
0.8232	2.420	0.5275	0.7471	2.061	0.5794
0.8230	2.419	0.5277	0.7449	2.053	0.5808
0.8227	2.417	0.5279	0.7428	2.045	0.5821
0.8224	2.416	0.5281	0.7406	2.038	0.5835
0.8222	2.414	0.5283	0.7385	2.030	0.5848
0.8219	2.412	0.5285	0.7363	2.022	0.5861
0.8216	2.411	0.5287	0.7342	2.015	0.5874
0.8214	2.409	0.5289	0.7321	2.008	0.5887
0.8211	2.408	0.5291	0.7300	2.000	0.5900
0.8208	2.406	0.5293	0.7279	1.993	0.5913
0.8206	2.404	0.5295	0.7258	1.986	0.5926
0.8203	2.403	0.5297	0.7238	1.979	0.5938
0.8200	2.401	0.5299	0.7217	1.973	0.5951
0.8198	2.400	0.5301	0.7197	1.966	0.5963
0.8195	2.398	0.5303	0.7176	1.959	0.5975
0.8192	2.397	0.5305	0.7156	1.953	0.5988
0.8190	2.395	0.5307	0.7136	1.947	0.6000
0.8187	2.393	0.5309	0.7116	1.940	0.6012
0.8184	2.392	0.5311	0.7096	1.934	0.6024
0.8182	2.390	0.5313	0.7076	1.928	0.6035
0.8179	2.389	0.5315	0.7056	1.922	0.6047
0.8176	2.387	0.5317	0.7036	1.916	0.6059
0.8174	2.386	0.5319	0.7017	1.910	0.6070
0.8171	2.384	0.5321	0.6997	1.904	0.6082
0.8168	2.383	0.5323	0.6978	1.899	0.6093
0.8166	2.381	0.5325	0.6959	1.893	0.6104
0.8139	2.366	0.5344	0.6940	1.887	0.6116
0.8113	2.352	0.5363	0.6920	1.882	0.6127
0.8087	2.338	0.5381	0.6901	1.876	0.6138
0.8062	2.324	0.5400	0.6882	1.871	0.6149
0.8036	2.310	0.5418	0.6864	1.866	0.6160
0.8011	2.297	0.5436	0.6845	1.861	0.6171
0.7986	2.284	0.5454	0.6826	1.855	0.6182
0.7961	2.272	0.5471	0.6807	1.850	0.6192
0.7936	3.259	0.5488	0.6789	1.845	0.6203
0.7911	2.248	0.5505	0.6771	1.840	0.6214
0.7887	2.236	0.5522	0.6752	1.835	0.6224
0.7862	2.224	0.5539	0.6734	1.831	0.6234
0.7838	2.213	0.5555	0.6716	1.826	0.6245
0.7814	2.202	0.5571	0.6698	1.821	0.6255
0.7790	2.191	0.5587	0.6680	1.816	0.6265
0.7767	2.181	0.5603	0.6662	1.812	0.6276
0.7743	2.171	0.5619	0.6644	1.807	0.6286
0.7720	2.161	0.5634	0.6626	1.803	0.6296
0.7696	2.151	0.5649	0.6608	1.798	0.6306
0.7673	2.141	0.5665	0.6590	1.794	0.6316
0.7650	2.131	0.5680	0.6573	1.789	0.6326
0.7627	2.122	0.5694	0.6555	1.785	0.6336
0.7605	2.113	0.5709	0.6538	1.781	0.6345
0.7582	2.104	0.5724	0.6521	1.776	0.6355
0.7560	2.095	0.5738	0.6503	1.772	0.6365
0.7537	2.087	0.5752	0.6486	1.768	0.6374
0.7515	2.078	0.5766	0.6469	1.764	0.6384

TABLE III-1 (Cont.)

COS (θ)	m	n	COS (θ)	m	n
0.6452	1.760	0.6394	0.5616	1.588	0.6847
0.6435	1.756	0.6403	0.5602	1.585	0.6854
0.6418	1.752	0.6413	0.5588	1.582	0.6862
0.6401	1.748	0.6422	0.5574	1.580	0.6869
0.6384	1.744	0.6431	0.5560	1.578	0.6876
0.6367	1.740	0.6440	0.5546	1.575	0.6884
0.6351	1.736	0.6450	0.5533	1.573	0.6891
0.6334	1.732	0.6459	0.5519	1.570	0.6898
0.6317	1.729	0.6468	0.5505	1.568	0.6906
0.6301	1.725	0.6477	0.5491	1.565	0.6913
0.6284	1.721	0.6486	0.5478	1.563	0.6920
0.6268	1.718	0.6495	0.5464	1.561	0.6927
0.6252	1.714	0.6504	0.5450	1.558	0.6934
0.6235	1.710	0.6513	0.5437	1.556	0.6942
0.6219	1.707	0.6522	0.5423	1.554	0.6949
0.6203	1.703	0.6531	0.5410	1.552	0.6956
0.6187	1.700	0.6540	0.5396	1.549	0.6963
0.6171	1.696	0.6548	0.5383	1.547	0.6970
0.6155	1.693	0.6557	0.5370	1.545	0.6977
0.6139	1.690	0.6566	0.5356	1.543	0.6984
0.6123	1.686	0.6574	0.5343	1.540	0.6991
0.6108	1.683	0.6583	0.5330	1.538	0.6998
0.6092	1.680	0.6591	0.5317	1.536	0.7005
0.6076	1.676	0.6600	0.5304	1.534	0.7012
0.6061	1.673	0.6608	0.5290	1.532	0.7019
0.6045	1.670	0.6617	0.5277	1.529	0.7026
0.6029	1.667	0.6625	0.5264	1.527	0.7032
0.6014	1.664	0.6634	0.5251	1.525	0.7039
0.5999	1.660	0.6642	0.5238	1.523	0.7046
0.5983	1.657	0.6650	0.5225	1.521	0.7053
0.5968	1.654	0.6658	0.5213	1.519	0.7060
0.5953	1.651	0.6667	0.5200	1.517	0.7066
0.5938	1.648	0.6675	0.5187	1.515	0.7073
0.5922	1.645	0.6683	0.5174	1.513	0.7080
0.5907	1.642	0.6691	0.5161	1.511	0.7086
0.5892	1.639	0.6699	0.5149	1.509	0.7093
0.5877	1.636	0.6707	0.5136	1.507	0.7100
0.5862	1.633	0.6715	0.5123	1.505	0.7106
0.5848	1.631	0.6723	0.5111	1.503	0.7113
0.5833	1.628	0.6731	0.5098	1.501	0.7120
0.5818	1.625	0.6739	0.5086	1.499	0.7126
0.5803	1.622	0.6747	0.5073	1.497	0.7133
0.5789	1.619	0.6755	0.5061	1.495	0.7139
0.5774	1.617	0.6763	0.5048	1.493	0.7146
0.5759	1.614	0.6770	0.5036	1.491	0.7152
0.5745	1.611	0.6778	0.5024	1.489	0.7159
0.5730	1.608	0.6786	0.5011	1.488	0.7165
0.5716	1.606	0.6794	0.4999	1.486	0.7171
0.5702	1.603	0.6801	0.4987	1.484	0.7178
0.5687	1.600	0.6809	0.4975	1.482	0.7184
0.5673	1.598	0.6816	0.4963	1.480	0.7191
0.5659	1.595	0.6824	0.4950	1.478	0.7197
0.5645	1.593	0.6832	0.4938	1.477	0.7203
0.5630	1.590	0.6839	0.4926	1.475	0.7210

TABLE III-1 (Cont.)

COS (θ)	m	n	COS (θ)	m	n
0.4914	1.473	0.7216	0.4308	1.389	0.7532
0.4902	1.471	0.7222	0.4297	1.388	0.7538
0.4890	1.469	0.7228	0.4287	1.386	0.7543
0.4878	1.468	0.6235	0.4276	1.385	0.7549
0.4866	1.466	0.7241	0.4266	1.384	0.7554
0.4854	1.464	0.7247	0.4255	1.383	0.7560
0.4843	1.462	0.6253	0.4245	1.381	0.7565
0.4831	1.461	0.7259	0.4235	1.380	0.7571
0.4819	1.459	0.7266	0.4224	1.379	0.7576
0.4807	1.457	0.7272	0.4214	1.377	0.7581
0.2795	1.456	0.7278	0.4204	1.376	0.7587
0.4784	1.454	0.7284	0.4193	1.375	0.7592
0.4772	1.452	0.7290	0.4183	1.373	0.7597
0.4760	1.451	0.7296	0.4173	1.372	0.7603
0.4749	1.449	0.7302	0.4163	1.371	0.7608
0.4737	1.447	0.7308	0.4153	1.370	0.7613
0.4726	1.446	0.7314	0.4142	1.368	0.7619
0.4714	1.444	0.7320	0.4132	1.367	0.7624
0.4703	1.442	0.7326	0.4122	1.366	0.7629
0.4691	1.441	0.7332	0.4112	1.365	0.7634
0.4680	1.439	0.7338	0.4102	1.363	0.7640
0.4668	1.438	0.7344	0.4092	1.362	0.7645
0.4657	1.436	0.7350	0.4082	1.361	0.7650
0.4646	1.434	0.7356	0.4072	1.360	0.7655
0.4634	1.433	0.7362	0.4062	1.358	0.7661
0.4623	1.431	0.7368	0.4052	1.357	0.7666
0.4612	1.430	0.7374	0.4042	1.356	0.7671
0.4601	1.428	0.7380	0.4032	1.355	0.7676
0.4589	1.427	0.7385	0.4023	1.354	0.7681
0.4578	1.425	0.7391	0.4013	1.352	0.7687
0.4567	1.424	0.7397	0.4003	1.351	0.7692
0.4556	1.422	0.7403	0.3993	1.350	0.7697
0.4545	1.421	0.7409	0.3983	1.349	0.7702
0.4534	1.419	0.7414	0.3974	1.348	0.7707
0.4523	1.418	0.7420	0.3964	1.347	0.7712
0.4512	1.416	0.7426	0.3954	1.345	0.7717
0.4501	1.415	0.7432	0.3944	1.344	0.7722
0.4490	1.413	0.7437	0.3935	1.343	0.7727
0.4479	1.412	0.7443	0.3925	1.342	0.7733
0.4468	1.410	0.7449	0.3916	1.341	0.7738
0.4457	1.409	0.7454	0.3906	1.340	0.7743
0.4446	1.407	0.7460	0.3896	1.339	0.7748
0.4435	1.406	0.7466	0.3887	1.337	0.7753
0.4425	1.404	0.7471	0.3877	1.336	0.7758
0.4414	1.403	0.7477	0.3868	1.335	0.7763
0.4403	1.402	0.7483	0.3858	1.334	0.7768
0.4392	1.400	0.7488	0.3849	1.333	0.7773
0.4382	1.399	0.7494	0.3839	1.332	0.7778
0.4371	1.397	0.7499	0.3830	1.331	0.7783
0.4360	1.396	0.7505	0.3820	1.330	0.7788
0.4350	1.395	0.7510	0.3811	1.329	0.7793
0.4339	1.393	0.7516	0.3802	1.328	0.7797
0.4329	1.392	0.7521	0.3792	1.326	0.7802
0.4318	1.391	0.7527	0.3783	1.325	0.7807

TABLE III-1 (Cont.)

COS (θ)	m	n	COS (θ)	m	n
0.3773	1.324	0.7812	0.3297	1.272	0.8065
0.3764	1.323	0.7817	0.3288	1.271	0.8069
0.3755	1.322	0.7822	0.3280	1.270	0.8074
0.3746	1.321	0.7827	0.3272	1.269	0.8078
0.3736	1.320	0.7832	0.3263	1.268	0.8083
0.3727	1.319	0.7837	0.3255	1.268	0.8087
0.3718	1.318	0.7842	0.3247	1.267	0.8091
0.3709	1.317	0.7846	0.3239	1.266	0.8096
0.3700	1.316	0.7851	0.3230	1.265	0.8100
0.3690	1.315	0.7856	0.3222	1.264	0.8105
0.3681	1.314	0.7861	0.3214	1.263	0.8109
0.3672	1.313	0.7866	0.3206	1.262	0.8113
0.3663	1.312	0.7870	0.3198	1.262	0.8118
0.3654	1.311	0.7875	0.3189	1.261	0.8122
0.3645	1.310	0.7880	0.3181	1.260	0.8127
0.3636	1.309	0.7885	0.3173	1.259	0.8131
0.3627	1.308	0.7890	0.3165	1.258	0.8135
0.3618	1.307	0.7894	0.3157	1.257	0.8140
0.3609	1.306	0.7899	0.3149	1.257	0.8144
0.3600	1.305	0.7904	0.3141	1.256	0.8148
0.3591	1.304	0.7909	0.3133	1.255	0.8153
0.3582	1.303	0.7913	0.3125	1.254	0.8157
0.3573	1.302	0.7918	0.3116	1.253	0.8161
0.3564	1.301	0.7923	0.3108	1.253	0.8166
0.3555	1.300	0.7927	0.3100	1.252	0.8170
0.3546	1.299	0.7932	0.3092	1.251	0.8174
0.3538	1.298	0.7937	0.3084	1.250	0.8178
0.3529	1.297	0.7941	0.3077	1.249	0.8183
0.3520	1.296	0.7946	0.3069	1.249	0.8187
0.3511	1.295	0.7951	0.3061	1.248	0.8191
0.3502	1.294	0.7955	0.3053	1.247	0.8196
0.3494	1.293	0.7960	0.3045	1.246	0.8200
0.3485	1.292	0.7965	0.3037	1.245	0.8204
0.3476	1.291	0.7969	0.3029	1.245	0.8308
0.3467	1.290	0.7974	0.3021	1.244	0.8213
0.3459	1.289	0.7979	0.3013	1.243	0.8217
0.3450	1.288	0.7983	0.3005	1.242	0.8221
0.3441	1.287	0.7988	0.2998	1.241	0.8225
0.3433	1.286	0.7992	0.2990	1.241	0.8229
0.3424	1.285	0.7997	0.2982	1.240	0.8234
0.3416	1.285	0.8002	0.2974	1.239	0.8238
0.3407	1.284	0.8006	0.2966	1.238	0.8242
0.3398	1.283	0.8011	0.2959	1.238	0.8246
0.3390	1.282	0.8015	0.2951	1.237	0.8250
0.3381	1.281	0.8020	0.2943	1.236	0.8255
0.3373	1.280	0.8024	0.2936	1.235	0.8259
0.3364	1.279	0.8029	0.2928	1.235	0.8263
0.3356	1.278	0.8033	0.2920	1.234	0.8267
0.3347	1.277	0.8038	0.2913	1.233	0.8271
0.3339	1.276	0.8042	0.2905	1.232	0.8275
0.3330	1.276	0.8047	0.2897	1.232	0.8280
0.3322	1.275	0.8051	0.2890	1.231	0.8284
0.3314	1.274	0.8056	0.2882	1.230	0.8288
0.3305	1.273	0.8060	0.2874	1.229	0.8292

TABLE III-1 (Cont.)

COS (θ)	m	n	COS (θ)	m	n
0.2867	1.229	0.8296	0.2475	1.192	0.8510
0.2859	1.228	0.8300	0.2469	1.191	0.8514
0.2852	1.227	0.8304	0.2462	1.190	0.8518
0.2844	1.226	0.8308	0.2455	1.190	0.8522
0.2837	1.226	0.8312	0.2448	1.189	0.8525
0.2829	1.225	0.8317	0.2441	1.189	0.8529
0.2822	1.224	0.8321	0.2434	1.188	0.8533
0.2814	1.223	0.8325	0.2427	1.187	0.8537
0.2807	1.223	0.8329	0.2420	1.187	0.8541
0.2799	1.222	0.8333	0.2414	1.186	0.8544
0.2792	1.221	0.8337	0.2407	1.185	0.8548
0.2784	1.221	0.8341	0.2400	1.185	0.8552
0.2777	1.220	0.8345	0.2393	1.184	0.8556
0.2769	1.219	0.8349	0.2386	1.184	0.8560
0.2762	1.218	0.8353	0.2380	1.183	0.8563
0.2754	1.218	0.8357	0.2373	1.182	0.8567
0.2747	1.217	0.8361	0.2366	1.182	0.8571
0.2740	1.216	0.8365	0.2359	1.181	0.8575
0.2732	1.216	0.8369	0.2353	1.181	0.8578
0.2725	1.215	0.8373	0.2346	1.180	0.8582
0.2718	1.214	0.8377	0.2339	1.179	0.8586
0.2710	1.214	0.8381	0.2332	1.179	0.8590
0.2704	1.213	0.8385	0.2326	1.178	0.8593
0.2696	1.212	0.8389	0.2319	1.178	0.8597
0.2688	1.211	0.8393	0.2312	1.177	0.8601
0.2681	1.211	0.8397	0.2306	1.176	0.8604
0.2674	1.210	0.8401	0.2299	1.176	0.8608
0.2667	1.209	0.8405	0.2292	1.175	0.8612
0.2659	1.209	0.8409	0.2286	1.175	0.8616
0.2652	1.208	0.8413	0.2279	1.174	0.8619
0.2645	1.207	0.8417	0.2273	1.173	0.8623
0.2638	1.207	0.8421	0.2266	1.173	0.8627
0.2631	1.206	0.8425	0.2259	1.172	0.8630
0.2623	1.205	0.8429	0.2253	1.172	0.8634
0.2616	1.205	0.8433	0.2246	1.171	0.8638
0.2609	1.204	0.8437	0.2240	1.171	0.8641
0.2602	1.203	0.8440	0.2233	1.170	0.8645
0.2595	1.203	0.8444	0.2227	1.169	0.8649
0.2588	1.202	0.8448	0.2220	1.169	0.8652
0.2581	1.201	0.8452	0.2214	1.168	0.8656
0.2574	1.201	0.8456	0.2207	1.168	0.8660
0.2567	1.200	0.8460	0.2201	1.167	0.8663
0.2559	1.199	0.8464	0.2194	1.167	0.8667
0.2552	1.199	0.8468	0.2188	1.166	0.8671
0.2545	1.198	0.8472	0.2181	1.165	0.8674
0.2538	1.197	0.8476	0.2175	1.165	0.8678
0.2531	1.197	0.8479	0.2168	1.164	0.8681
0.2524	1.196	0.8483	0.2162	1.164	0.8685
0.2517	1.195	0.8487	0.2155	1.163	0.8689
0.2510	1.195	0.8491	0.2149	1.163	0.8692
0.2503	1.194	0.8495	0.2143	1.162	0.8696
0.2496	1.194	0.8499	0.2136	1.161	0.8699
0.2489	1.193	0.8503	0.2130	1.161	0.8703
0.2482	1.192	0.8506	0.2123	1.160	0.8707

TABLE III-1 (Cont.)

COS (θ)	m	n	COS (θ)	m	n
0.2117	1.160	0.8710	0.1786	1.132	0.8898
0.2111	1.159	0.8714	0.1781	1.131	0.8902
0.2104	1.159	0.8717	0.1775	1.131	0.8905
0.2098	1.158	0.8721	0.1769	1.130	0.8909
0.2092	1.158	0.8725	0.1763	1.130	0.8912
0.2085	1.157	0.8728	0.1757	1.130	0.8915
0.2079	1.157	0.8732	0.1751	1.129	0.8919
0.2073	1.156	0.8735	0.1745	1.129	0.8922
0.2066	1.155	0.8739	0.1740	1.128	0.8925
0.2060	1.155	0.8742	0.1734	1.128	0.8929
0.2054	1.154	0.8746	0.1728	1.127	0.8932
0.2047	1.154	0.8749	0.1722	1.127	0.8935
0.2041	1.153	0.8753	0.1716	1.126	0.8939
0.2035	1.153	0.8757	0.1711	1.126	0.8942
0.2029	1.152	0.8760	0.1705	1.125	0.8945
0.2023	1.152	0.8764	0.1699	1.125	0.8949
0.2016	1.151	0.8767	0.1693	1.124	0.8952
0.2010	1.151	0.8771	0.1688	1.124	0.8955
0.2004	1.150	0.8774	0.1682	1.123	0.8959
0.1998	1.150	0.8778	0.1676	1.123	0.8962
0.1992	1.149	0.8781	0.1670	1.122	0.8965
0.1985	1.149	0.8785	0.1665	1.122	0.8969
0.1979	1.148	0.8788	0.1659	1.122	0.8972
0.1973	1.148	0.8792	0.1653	1.121	0.8975
0.1967	1.147	0.8795	0.1648	1.121	0.8979
0.1961	1.146	0.8799	0.1642	1.120	0.8982
0.1955	1.146	0.8802	0.1636	1.120	0.8985
0.1948	1.145	0.8806	0.1631	1.119	0.8989
0.1942	1.145	0.8809	0.1625	1.119	0.8992
0.1936	1.144	0.8813	0.1619	1.118	0.8995
0.1930	1.144	0.8816	0.1614	1.118	0.8998
0.1924	1.143	0.8820	0.1608	1.117	0.9002
0.1918	1.143	0.8823	0.1602	1.117	0.9005
0.1912	1.142	0.8826	0.1597	1.116	0.9008
0.1906	1.142	0.8830	0.1591	1.116	0.9012
0.1900	1.141	0.8833	0.1585	1.116	0.9015
0.1894	1.141	0.8837	0.1580	1.115	0.9018
0.1888	1.140	0.8840	0.1574	1.115	0.9021
0.1882	1.140	0.8844	0.1569	1.114	0.9025
0.1876	1.139	0.8847	0.1563	1.114	0.9028
0.1870	1.139	0.8851	0.1558	1.113	0.9031
0.1864	1.138	0.8854	0.1552	1.113	0.9034
0.1858	1.138	0.8857	0.1546	1.112	0.9038
0.1852	1.137	0.8861	0.1541	1.112	0.9041
0.1846	1.137	0.8864	0.1535	1.112	0.9044
0.1840	1.136	0.8868	0.1530	1.111	0.9047
0.1834	1.136	0.8871	0.1524	1.111	0.9051
0.1828	1.135	0.8875	0.1519	1.110	0.9054
0.1822	1.135	0.8878	0.1513	1.110	0.9057
0.1816	1.134	0.8881	0.1508	1.109	0.9060
0.1810	1.134	0.8885	0.1502	1.109	0.9063
0.1804	1.133	0.8888	0.1497	1.109	0.9067
0.1798	1.133	0.8892	0.1491	1.108	0.9070
0.1792	1.132	0.8895	0.1486	1.108	0.9073

TABLE III-1 (Cont.)

COS (θ)	m	n	COS (θ)	m	n
0.1480	1.107	0.9076	0.1200	1.085	0.9242
0.1475	1.107	0.9080	0.1195	1.085	0.9245
0.1469	1.106	0.9083	0.1190	1.085	0.9249
0.1464	1.106	0.9086	0.1185	1.084	0.9252
0.1458	1.105	0.9089	0.1180	1.084	0.9255
0.1453	1.105	0.9092	0.1175	1.084	0.9258
0.1448	1.105	0.9096	0.1170	1.083	0.9261
0.1442	1.104	0.9099	0.1165	1.083	0.9264
0.1437	1.104	0.9102	0.1160	1.082	0.9267
0.1431	1.103	0.9105	0.1155	1.082	0.9270
0.1426	1.103	0.9108	0.1150	1.082	0.9273
0.1421	1.102	0.9111	0.1145	1.081	0.9276
0.1415	1.102	0.9115	0.1140	1.081	0.9279
0.1410	1.102	0.9118	0.1135	1.080	0.9282
0.1404	1.101	0.9121	0.1130	1.080	0.9285
0.1399	1.101	0.9124	0.1125	1.080	0.9288
0.1394	1.100	0.9127	0.1120	1.079	0.9291
0.1388	1.100	0.9130	0.1115	1.079	0.9294
0.1383	1.100	0.9134	0.1110	1.079	0.9297
0.1378	1.099	0.9137	0.1105	1.078	0.9300
0.1372	1.099	0.9140	0.1100	1.078	0.9303
0.1367	1.099	0.9143	0.1095	1.077	0.9306
0.1362	1.098	0.9146	0.1090	1.077	0.9309
0.1356	1.097	0.9149	0.1085	1.077	0.9312
0.1351	1.097	0.9153	0.1080	1.076	0.9315
0.1346	1.097	0.9156	0.1075	1.076	0.9318
0.1341	1.096	0.9159	0.1070	1.076	0.9321
0.1335	1.096	0.9162	0.1065	1.075	0.9324
0.1330	1.095	0.9165	0.1060	1.075	0.9327
0.1325	1.095	0.9168	0.1055	1.074	0.9330
0.1320	1.095	0.9171	0.1050	1.074	0.9333
0.1314	1.094	0.9174	0.1045	1.074	0.9336
0.1309	1.094	0.9178	0.1040	1.073	0.9339
0.1304	1.093	0.9181	0:1036	1.073	0.9342
0.1299	1.093	0.9184	0.1031	1.073	0.9345
0.1293	1.093	0.9187	0.1026	1.072	0.9348
0.1288	1.092	0.9190	0.1021	1.072	0.9351
0.1283	1.092	0.9193	0.1016	1.072	0.9354
0.1278	1.091	0.9196	0.1011	1.071	0.9357
0.1273	1.091	0.9199	0.1006	1.071	0.9360
0.1267	1.091	0.9202	0.1001	1.070	0.9363
0.1262	1.090	0.9205	0.0997	1.070	0.9366
0.1257	1.090	0.9209	0.0992	1.070	0.9369
0.1252	1.089	0.9212	0.0987	1.069	0.9372
0.1247	1.089	0.9215	0.0982	1.069	0.9375
0.1241	1.089	0.9218	0.0977	1.069	0.9377
0.1236	1.088	0.9221	0.0972	1.068	0.9380
0.1231	1.088	0.9224	0.0968	1.068	0.9383
0.1226	1.087	0.9227	0.0963	1.068	0.9386
0.1221	1.087	0.9230	0.0958	1.067	0.9389
0.1216	1.087	0.9233	0.0953	1.067	0.9392
0.1211	1.086	0.9236	0.0948	1.067	0.9395
0.1206	1.086	0.9239	0.0944	1.066	0.9398
			0.0939	1.066	0.9401

TABLE III-1 (Cont.)

COS (θ)	m	n	COS (θ)	m	n
0.0934	1.065	0.9404	0.0684	1.047	0.9558
0.0929	1.065	0.9407	0.0680	1.047	0.9561
0.0924	1.065	0.9410	0.0675	1.047	0.9564
0.0920	1.064	0.9413	0.0671	1.046	0.9567
0.0915	1.064	0.9416	0.0666	1.046	0.9570
0.0910	1.064	0.9418	0.0662	1.046	0.9572
0.0905	1.063	0.9421	0.0657	1.045	0.9575
0.0901	1.063	0.9424	0.0653	1.045	0.9578
0.0896	1.063	0.9427	0.0649	1.045	0.9581
0.0801	1.062	0.9430	0.0644	1.044	0.9584
0.0887	1.062	0.9433	0.0640	1.044	0.9586
0.0882	1.062	0.9436	0.0635	1.044	0.9589
0.0877	1.061	0.9439	0.0632	1.043	0.9592
0.0872	1.061	0.9442	0.0626	1.043	0.9595
0.0868	1.061	0.9445	0.0622	1.043	0.9597
0.0863	1.060	0.9447	0.0618	1.043	0.9600
0.0858	1.060	0.9450	0.0613	1.042	0.9603
0.0854	1.060	0.9453	0.0609	1.042	0.9606
0.0849	1.059	0.9456	0.0604	1.042	0.9609
0.0844	1.059	0.9459	0.0600	1.041	0.9611
0.0840	1.059	0.9462	0.0596	1.041	0.9614
0.0835	1.058	0.9465	0.0591	1.041	0.9617
0.0830	1.058	0.9468	0.0587	1.040	0.9620
0.0826	1.058	0.9471	0.0582	1.040	0.9622
0.0821	1.057	0.9473	0.0578	1.040	0.9625
0.0816	1.057	0.9476	0.0574	1.039	0.9628
0.0812	1.057	0.9479	0.0569	1.039	0.9631
0.0807	1.056	0.9482	0.0565	1.039	0.9633
0.0802	1.056	0.9485	0.0561	1.038	0.9636
0.0798	1.055	0.9488	0.0556	1.038	0.9639
0.0793	1.055	0.9491	0.0552	1.038	0.9642
0.0789	1.055	0.9493	0.0548	1.038	0.9644
0.0784	1.054	0.9496	0.0543	1.037	0.9647
0.0779	1.054	0.9499	0.0539	1.037	0.9650
0.0775	1.054	0.9502	0.0535	1.037	0.9653
0.0770	1.053	0.9505	0.0530	1.036	0.9655
0.0766	1.053	0.9508	0.0526	1.036	0.9658
0.0761	1.053	0.9510	0.0522	1.036	0.9661
0.0757	1.052	0.9513	0.0518	1.035	0.9663
0.0752	1.052	0.9516	0.0513	1.035	0.9666
0.0747	1.052	0.9519	0.0509	1.035	0.9669
0.0743	1.052	0.9522	0.0505	1.035	0.9672
0.0738	1.051	0.9525	0.0500	1.034	0.9674
0.0734	1.051	0.9527	0.0496	1.034	0.9677
0.0729	1.051	0.9530	0.0492	1.034	0.9680
0.0725	1.050	0.9533	0.0488	1.033	0.9683
0.0720	1.050	0.9536	0.0483	1.033	0.9685
0.0716	1.050	0.9539	0.0479	1.033	0.9688
0.0711	1.049	0.9542	0.0475	1.032	0.9691
0.0707	1.049	0.9544	0.0471	1.032	0.9693
0.0702	1.049	0.9547	0.0466	1.032	0.9696
0.0698	1.050	0.9550	0.0462	1.032	0.9699
0.0693	1.048	0.9553	0.0458	1.031	0.9701
0.0689	1.048	0.9556	0.0454	1.031	0.9704

TABLE   III-1 (Cont.)

COS (θ)	m	n	COS (θ)	m	n
0.0449	1.031	0.9707	0.0216	1.015	0.9857
0.0445	1.030	0.9710	0.0212	1.014	0.9860
0.0441	1.030	0.9712	0.0208	1.014	0.9863
0.0437	1.030	0.9715	0.0204	1.014	0.9865
0.0433	1.029	0.9718	0.0200	1.013	0.9868
0.0428	1.029	0.9720	0.0196	1.013	0.9870
0.0424	1.029	0.9723	0.0192	1.013	0.9873
0.0420	1.029	0.9726	0.0188	1.013	0.9876
0.0416	1.028	0.9728	0.0184	1.012	0.9878
0.0412	1.028	0.9731	0.0181	1.012	0.9881
0.0407	1.028	0.9734	0.0177	1.012	0.9883
0.0403	1.027	0.9736	0.0173	1.012	0.9886
0.0399	1.027	0.9739	0.0169	1.011	0.9888
0.0395	1.027	0.9742	0.0165	1.011	0.9891
0.0391	1.027	0.9744	0.0161	1.011	0.9894
0.0387	1.026	0.9747	0.0157	1.011	0.9896
0.0382	1.026	0.9750	0.0153	1.010	0.9899
0.0378	1.026	0.9752	0.0149	1.010	0.9901
0.0374	1.025	0.9755	0.0145	1.010	0.9904
0.0370	1.025	0.9758	0.0141	1.009	0.9906
0.0366	1.025	0.9760	0.0137	1.009	0.9909
0.0362	1.025	0.9763	0.0134	1.009	0.9912
0.0358	1.024	0.9766	0.0130	1.009	0.9914
0.0354	1.024	0.9768	0.0126	1.008	0.9917
0.0349	1.024	0.9771	0.0122	1.008	0.9919
0.0345	1.023	0.9774	0.0118	1.008	0.9922
0.0341	1.023	0.9776	0.0114	1.008	0.9924
0.0337	1.023	0.9779	0.0110	1.007	0.9927
0.0333	1.023	0.9782	0.0106	1.007	0.9929
0.0329	1.022	0.9784	0.0103	1.007	0.9932
0.0325	1.022	0.9787	0.0099	1.007	0.9934
0.0321	1.022	0.9789	0.0095	1.006	0.9937
0.0317	1.021	0.9792	0.0091	1.006	0.9940
0.0313	1.021	0.9795	0.0087	1.006	0.9942
0.0309	1.023	0.9797	0.0083	1.006	0.9945
0.0304	1.021	0.9800	0.0080	1.005	0.9947
0.0300	1.020	0.9803	0.0076	1.005	0.9950
0.0296	1.020	0.9805	0.0072	1.005	0.9952
0.0292	1.020	0.9808	0.0068	1.005	0.9955
0.0288	1.020	0.9811	0.0064	1.004	0.9957
0.0284	1.019	0.9813	0.0060	1.004	0.9960
0.0280	1.019	0.9816	0.0057	1.004	0.9962
0.0276	1.019	0.9818	0.0053	1.004	0.9965
0.0272	1.018	0.9821	0.0049	1.003	0.9967
0.0268	1.018	0.9824	0.0045	1.003	0.9970
0.0264	1.018	0.9826	0.0041	1.003	0.9972
0.0260	1.018	0.9829	0.0038	1.003	0.9975
0.0256	1.017	0.9831	0.0034	1.002	0.9977
0.0252	1.017	0.9834	0.0030	1.002	0.9980
0.0248	1.017	0.9837	0.0026	1.002	0.9982
0.0244	1.016	0.9839	0.0023	1.002	0.9985
0.0240	1.016	0.9842	0.0019	1.001	0.9987
0.0236	1.016	0.9844	0.0015	1.001	0.9990
0.0232	1.016	0.9847	0.0011	1.001	0.9992
0.0228	1.015	0.9850	0.0008	1.001	0.9995
0.0224	1.015	0.9852	0.0004	1.000	0.9997
0.0220	1.015	0.9855	0.0000	1.000	1.0000

TABLE III-2
Young's Modulus and Poisson's Ratio

Material	Young's Modulus (E) (psi)	Poisson's Ratio ν
Aluminum and Aluminum Alloys	10×10^6	0.33
Anodized Aluminum	10×10^6	0.33
Beryllium	$40 - 44 \times 10^6$	0.024 – 0.030
Beryllium Copper	19×10^6	0.34
Brass	16×10^6	0.33
Cadmium	10×10^6	0.44
Cast Iron	$14 - 21 \times 10^6$	0.21 – .30
Chromium	36×10^6	0.30
Copper	$16 - 17 \times 10^6$	0.33 – .35
Copper-Nickel	10×10^6	0.34
Delrin	4.74×10^5	0.35
Gold	10×10^6	0.42
HyMu' 80	30×10^6	0.30
Invar	21.4×10^6	0.26
Iron	30×10^6	0.30
Lead	2.2×10^6	0.41
Magnesium	6×10^6	0.25
Molybdenum	47×10^6	0.307
Monel	24.4×10^6	0.25 – .32
Nickel	30×10^6	0.41
Nylatron G	4.56×10^5	0.40
Nylatron GS	4.75×10^5	0.40
Phosphorous Bronze	16×10^6	0.34
Platinum	24×10^6	0.39
Polyethylene 1	2.43×10^4	0.35
Polyethylene 2	6.16×10^4	0.35
Polyethylene 3	1.38×10^5	0.35
Polystyrene	3.68×10^5	0.35
Silver	10×10^6	0.37
Sintered Brass	16×10^6	0.34
Sintered Bronze	16×10^6	0.34
Sintered Iron	30×10^6	0.30
Sintered Iron-Copper	30×10^6	0.30
Sintered Steel	30×10^6	0.30
Stainless Steel 18 - 8	27×10^6	0.30
Steel	30×10^6	0.30
Teflon	5.7×10^4	0.35
Tungsten	51×10^6	0.19
Zinc	13×10^6	0.28
Zytel 101	4.65×10^5	0.40

TABLE III-3

γ_R For Various Combinations

52100/Stainless Steel					52100/Steel			
Material	Lubrication	γ_R	μ		Material	Lubrication	γ_R	μ
302	Dry	0.2	1.00		1045	Dry	0.20	0.67
	A	0.2	0.19			A	0.54	0.15
	B	0.2	0.16			B	0.20	0.17
	C	0.2	0.21			C	0.20	0.28
						D	0.54	0.08
303 E Z	Dry	0.2	0.79					
	A	0.2	0.21		1055	Dry	0.20	0.74
	B	0.2	0.19			A	0.20	0.18
	C	0.2	0.32			B	0.20	0.14
						C	0.20	0.23
321	Dry	0.20	1.16			D	0.54	0.09
	A	0.54	0.17					
	B	0.54	0.13		1060	Dry	0.20	0.73
	C	0.20	0.18			A	0.20	0.14
	D	0.54	0.08			B	0.20	0.21
						C	0.20	0.31
347	Dry	0.2	1.15			D	0.54	0.16
	A	0.2	0.16					
	B	0.2	0.15		1085	Dry	0.20	0.81
	C	0.2	0.12			A	0.20	0.20
	D	0.54	0.11			B	0.54	0.14
						C	0.20	0.22
410	Dry	0.20	0.85			D	0.54	0.08
	A	0.54	0.15					
	B	0.54	0.15		4140 L L	Dry	0.20	0.57
	C	0.20	0.20			A	0.20	0.21
	D	0.54	0.08			B	0.20	0.17
						C	0.20	0.26
416 E Z	Dry	0.2	0.97			D	0.54	0.10
	A	0.2	0.13					
	B	0.54	0.15		4150	Dry	0.20	0.67
	C	0.20	0.23			A	0.20	0.15
						B	0.20	0.15
440 C	Dry	0.20	0.66			C	0.20	0.23
	A	0.20	0.18					
	B	0.20	0.13		4620	Dry	0.20	0.79
	C	0.20	0.17			A	0.54	0.19
	D	0.54	0.08			B	0.54	0.15
52100/Steel						C	0.20	0.27
Material	Lubrication	γ_R	μ					
1018	Dry	0.20	0.80					
	A	0.54	0.18					
	B	0.54	0.21					
	C	0.20	0.26					

TABLE III-3 (Cont.)

52100/Steel

Material	Lubrication	γ_R	μ
5130LL	Dry	0.20	0.62
	A	0.20	0.16
	B	0.20	0.16
	C	0.20	0.21
	D	0.54	0.17
8214	Dry	0.20	0.71
	A	0.20	0.17
	B	0.54	0.19
	C	0.54	0.20
	D	0.54	0.09
8620	Dry	0.20	0.70
	A	0.20	0.22
	B	0.54	0.17
	C	0.20	0.25
52100	Dry	0.20	0.60
	A	0.20	0.21
	B	0.54	0.16
	C	0.20	0.21
	D	0.54	0.10
Carpenter 11, Special Steel Annealed Water Hard	Dry	0.20	0.78
	A	0.54	0.18
	B	0.54	0.16
	C	0.54	0.21
	D	0.54	0.09
Hampden Steel Annealed Oil Wear	Dry	0.54	
	A	0.54	0.13
	B	0.54	0.12
	C	0.54	0.11
	D	0.54	0.08
HYCC (HA)	Dry	0.20	0.62
	A	0.54	0.13
	B	0.54	0.11
	C	0.20	0.15
	D	0.54	0.08

52100/Steel

Material	Lubrication	γ_R	μ
HYCC (PM)	Dry	0.20	0.64
	A	0.20	0.16
	B	0.20	0.17
	C	0.20	0.32
Ketos	Dry	0.20	0.67
	A	0.54	0.18
	B	0.54	0.15
	C	0.20	0.31
Nitralloy-G	Dry	0.20	0.63
	A	0.20	0.15
	B	0.20	0.13
	C	0.20	0.11
	D	0.54	0.09
Rexalloy AA	Dry	0.20	0.73
	A	0.54	0.13
	B	0.20	0.13
	C	0.20	0.14
Star Zenith Steel Annealed Red Wear	Dry	0.20	0.63
	A	0.54	0.12
	B	0.20	0.12
	C	0.54	0.16
	D	0.54	0.09

52100/Nickel Alloy

Material	Lubrication	γ_R	μ
Carpenter Free Cut Invar "36" Annealed	Dry	0.20	1.28
	A	0.20	0.24
	B	0.20	0.18
	C	0.20	0.26
Hy Mu 80 Annealed	Dry	0.20	1.00
	A	0.20	0.16
	B	0.54	0.13
	C	0.20	0.16
Monel C	Dry	0.20	0.73
	A	0.20	0.12
	B	0.54	0.14
	C	0.20	0.17

TABLE III-3 (Cont.)

52100/Copper Alloy

Material	Lubrication	γ_R	μ
Be-Cu	Dry	0.20	0.78
	A	0.54	0.29
	B	0.20	0.15
	C	0.54	0.29
	D	0.54	0.35
Cu-Ni	Dry	0.20	1.23
	A	0.54	0.21
	B	0.54	0.15
	C	0.20	0.26
	D	0.54	0.10
Phos-Bronze A	Dry	0.20	0.67
	A	0.20	0.19
	B	0.54	0.16
	C	0.54	0.18
	D	0.54	0.12

52100/Aluminum Alloy

Material	Lubrication	γ_R	μ
43 Aluminum	Dry	0.20	1.42
	A	0.54	0.12
	B	0.54	0.22
	C	0.54	0.25
	D	0.54	0.10
112 Aluminum	Dry	0.20	1.08
	A	0.54	0.25
	B	0.20	0.15
	C	0.20	0.19
	D	0.54	0.10
195 Aluminum	Dry	0.20	1.07
	A	0.54	0.17
	B	0.54	0.13
	C	0.54	0.19
	D	0.54	0.10
220 Aluminum	Dry	0.20	0.79
	A	0.54	0.25
	B	0.20	0.23
	C	0.20	0.24
	D	0.54	0.15

52100/Aluminum Alloy

Material	Lubrication	γ_R	μ
355 Aluminum	Dry	0.20	1.21
	A	0.54	0.13
	B	0.54	0.20
	C	0.20	0.24
	D	0.54	0.09
356 Aluminum	Dry	0.20	1.40
	A	0.54	0.22
	B	0.54	0.17
	C	0.54	0.23
	D	0.54	0.10

52100/Sintered Brass

Material	Lubrication	γ_R	μ
Sintered Brass 1	Dry	0.20	0.32
	A	0.20	0.21
	B	0.20	0.16
	C	0.20	0.23
Sintered Brass 2	Dry	0.54	0.37
	A	0.54	0.33
	B	0.54	0.20
	C	0.54	0.31

52100/Sintered Bronze

Material	Lubrication	γ_R	μ
Sintered Bronze 1	Dry	0.20	0.26
	A	0.20	0.23
	B	0.20	0.11
	C	0.20	0.18
Sintered Bronze 2	Dry	0.20	0.31
	A	0.20	0.17
	B	0.20	0.23
	C	0.20	0.25

TABLE III-3 (Cont.)

52100/Sintered Iron

Material	Lubrication	γ_R	μ
Sintered Iron 1	Dry	0.20	0.38
	A	0.20	0.21
	B	0.54	0.23
	C	0.54	0.24
Sintered Iron 2	Dry	0.54	0.40
	A	0.54	0.26
	B	0.54	0.23
	C	0.54	0.26
Sintered Iron 3	Dry	0.54	0.45
	A	0.54	0.23
	B	0.54	0.24
	C	0.54	0.24

52100/Sintered Iron-Copper

Material	Lubrication	γ_R	μ
Sintered Iron-Copper 1	Dry	0.20	0.47
	A	0.20	0.20
	B	0.54	0.19
	C	0.54	0.19
Sintered Iron-Copper 2	Dry	0.20	0.43
	A	0.20	0.21
	B	0.54	0.26
	C	0.54	0.27

52100/Sintered Steel

Material	Lubrication	γ_R	μ
Sintered Steel 1	Dry	0.20	0.34
	A	0.54	0.15
	B	0.54	0.15
	C	0.20	0.14
Sintered Steel 2	Dry	0.20	0.33
	A	0.54	0.21
	B	0.54	0.20
	C	0.54	0.25

52100/Sintered Steel

Material	Lubrication	γ_R	μ
Sintered Steel 3	Dry	0.20	0.50
	A	0.54	0.25
	B	0.54	0.16
	C	0.54	0.23

52100/Layered Material

Material	Lubrication	γ_R	μ
Plating 1	Dry	0.20	0.64
Plating 2	Dry	0.20	0.69
Plating 3	Dry	0.20	0.69
Plating 4	Dry	0.20	0.69
Nickel-Cobalt Plating 1	Dry	0.20	0.38
Nickel-Cobalt Plating 2	Dry	0.54	0.48
Chromium Plating	Dry	0.20	0.51
Anodized Aluminum	Dry	0.54	0.16

302/Stainless Steel

Material	Lubrication	γ_R	μ
302	Dry	0.20	1.02
	A	0.20	0.16
	B	0.20	0.15
	C	0.20	0.17
303 E Z	Dry	0.20	0.86
	A	0.54	0.16
	B	0.54	0.15
	C	0.20	0.14
321	Dry	0.20	1.47
	A	0.54	0.15
	B	0.54	0.14
	C	0.54	0.17
	D	0.54	0.08

TABLE III-3 (Cont.)

302/Stainless Steel				302/Steel			
Material	Lubrication	γ_R	μ	Material	Lubrication	γ_R	μ
347	Dry	0.20	1.33	1060	Dry	0.20	0.88
	A	0.54	0.15		A	0.54	0.16
	B	0.54	0.15		B	0.20	0.15
	C	0.54	0.16		C	0.20	0.16
	D	0.54	0.12		D	0.54	0.08
410	Dry	0.20	1.00	1085	Dry	0.20	1.03
	A	0.54	0.16		A	0.20	0.18
	B	0.54	0.14		B	0.54	0.15
	C	0.20	0.16		C	0.20	0.16
416EZ	Dry	0.20	0.87		D	0.54	0.09
	A	0.54	0.14	4140LL	Dry	0.20	0.78
	B	0.54	0.15		A	0.54	0.14
	C	0.20	0.21		B	0.54	0.14
	D	0.54	0.08		C	0.20	0.15
440 C	Dry	0.20	0.90		D	0.54	0.09
	A	0.54	0.13	4150	Dry	0.20	0.90
	B	0.20	0.15		A	0.54	0.17
	C	0.54	0.15		B	0.20	0.15
	D	0.54	0.11		C	0.20	0.20
					D	0.54	0.08

302/Steel							
Material	Lubrication	γ_R	μ	4620	Dry	0.20	0.96
1018	Dry	0.20	0.80		A	0.20	0.18
	A	0.54	0.14		B	0.54	0.15
	B	0.54	0.16		C	0.20	0.17
	C	0.54	0.18		D	0.54	0.08
1045	Dry	0.20	0.71	5130LL	Dry	0.20	0.84
	A	0.20	0.16		A	0.20	0.16
	B	0.54	0.14		B	0.20	0.14
	C	0.54	0.15		C	0.20	0.17
	D	0.54	0.11		D	0.54	0.08
1055	Dry	0.20	0.90	8214	Dry	0.20	0.86
	A	0.20	0.17		A	0.54	0.16
	B	0.54	0.16		B	0.54	0.15
	C	0.20	0.13		C	0.20	0.19
					D	0.54	0.08

TABLE III-3 (Cont.)

302/Steel					302/Nickel Alloy			
Material	Lubrication	γ_R	μ		Material	Lubrication	γ_R	μ
8620	Dry	0.20	0.83			Dry	0.20	1.33
	A	0.20	0.21		Carpenter Free	A	0.20	0.16
	B	0.20	0.15		Cut Invar "36"	B	0.20	0.19
	C	0.20	0.19		Annealed	C	0.20	0.24
Carpenter 11	Dry	0.20	0.84		Hy-Mu 80	Dry	0.20	1.18
Special Steel	A	0.54	0.16		Annealed	A	0.20	0.16
Annealed	B	0.20	0.14			B	0.20	0.17
Water Hard	C	0.20	0.16			C	0.20	0.18
Hampdin Steel	Dry	0.20	0.79		Monel C	Dry	0.20	0.99
Annealed Oil	A	0.54	0.16			A	0.20	0.15
Wear	B	0.54	0.13			B	0.20	0.15
	C	0.54	0.14			C	0.20	0.14
	D	0.54	0.13		302/Copper Alloy			
HYCC (HA)	Dry	0.54	0.89		Material	Lubrication	γ_R	μ
	A	0.54	0.14		Be-Cu	Dry	0.20	0.92
	B	0.54	0.14			A	0.54	0.22
	C	0.54	0.16			B	0.20	0.12
	D	0.54	0.12			C	0.54	0.16
HYCC (PM)	Dry	0.20	1.00			D	0.54	0.32
	A	0.20	0.17		Cu-Ni	Dry	0.20	1.17
	B	0.54	0.15			A	0.54	0.14
	C	0.20	0.16			B	0.54	0.17
Ketos	Dry	0.20	0.95			C	0.54	0.19
	A	0.54	0.16		Phos-Bronze A	Dry	0.20	0.74
Nitralloy G	Dry	0.20	0.83			A	0.20	0.15
	A	0.54	0.14			B	0.20	0.17
	B	0.54	0.14			C	0.20	0.15
	C	0.54	0.16			D	0.54	0.09
	D	0.54	0.09		302/Aluminum Alloy			
Rexalloy AA	Dry	0.20	1.03		Material	Lubrication	γ_R	μ
	A	0.54	0.15		43 Aluminum	Dry	0.20	1.67
	B	0.54	0.15			A	0.54	0.24
	C	0.54	0.15			B	0.54	0.24
	D	0.54	0.12			C	0.54	0.18
Star Zenith	Dry	0.20	0.93			D	0.54	0.08
Steel Annealed	A	0.54	0.15		112 Aluminum	Dry	0.20	1.16
Red Wear	B	0.54	0.14			A	0.54	0.20
	C	0.54	0.15			B	0.54	0.14
	D	0.54	0.11			C	0.54	0.13
						D	0.54	0.11

TABLE III-3 (Cont.)

302/Aluminum Alloy

Material	Lubrication	γ_R	μ
195 Aluminum	Dry	0.20	1.17
	A	0.54	0.15
	B	0.54	0.14
	C	0.54	0.20
	D	0.54	0.10
220 Aluminum	Dry	0.20	0.92
	A	0.20	0.14
	B	0.54	0.17
	C	0.20	0.25
	D	0.54	0.13
355 Aluminum	Dry	0.20	1.11
	A	0.54	0.17
	B	0.54	0.20
	C	0.54	0.18
	D	0.54	0.12
356 Aluminum	Dry	0.20	1.78
	A	0.54	0.18
	B	0.54	0.21
	C	0.54	0.18
	D	0.54	0.10

302/Plastic

Material	Lubrication	γ_R	μ
Delrin	Dry	0.54	0.36
	A	0.54	0.15
	B	0.54	0.18
	C	0.54	0.17
Nylatron G	Dry	0.54	0.57
	A	0.54	0.22
	B	0.54	0.24
	C	0.54	0.22
Nylatron GS	Dry	0.54	0.60
	A	0.54	0.24
	B	0.54	0.24
	C	0.54	0.24
Polyethylene 1	Dry	0.54	0.26
	A	0.54	0.17
	B	0.54	0.17
	C	0.54	0.13

302/Plastic

Material	Lubrication	γ_R	μ
Polyethylene 2	Dry	0.54	0.31
	A	0.54	0.22
	B	0.54	0.26
	C	0.54	0.20
Polyethylene 3	Dry	0.54	0.40
	A	0.54	0.24
	B	0.54	0.30
	C	0.54	0.26
Polystyrene	Dry	0.54	0.60
	A	0.54	0.30
	B	0.54	0.31
	C	0.54	0.31
Teflon	Dry	0.54	0.09
	A	0.54	0.15
	B	0.54	0.11
	C	0.54	0.12
Zytel 101	Dry	0.54	0.60
	A	0.54	0.27
	B	0.54	0.27
	C	0.54	0.23

Brass/Stainless Steel

Material	Lubrication	γ_R	μ
302	Dry	0.20	0.70
	A	0.20	0.22
	B	0.54	0.19
	C	0.20	0.18
303 EZ	Dry	0.20	0.77
	A	0.20	0.22
	B	0.54	0.19
	C	0.20	0.22
	D	0.54	0.14
321	Dry	0.20	0.78
	A	0.20	0.23
	B	0.54	0.13
	C	0.20	0.23
347	Dry	0.20	0.82
	A	0.20	0.19
	B	0.20	0.23
	C	0.20	0.22

TABLE III-3 (Cont.)

Material	Brass/Stainless Steel Lubrication	γ_R	μ
410	Dry	0.20	0.67
	A	0.20	0.23
	B	0.20	0.17
	C	0.20	0.21
416 EZ	Dry	0.20	0.66
	A	0.20	0.18
	B	0.20	0.18
	C	0.20	0.18
440 C	Dry	0.20	0.72
	A	0.20	0.18
	B	0.54	0.16
	C	0.20	0.18

Material	Brass/Steel Lubrication	γ_R	μ
1018	Dry	0.20	0.60
	A	0.20	0.24
	B	0.54	0.20
	C	0.20	0.28
	D	0.54	0.20
1045	Dry	0.20	0.66
	A	0.20	0.20
	B	0.20	0.12
	C	0.20	0.25
1055	Dry	0.20	0.69
	A	0.20	0.20
	B	0.20	0.22
	C	0.20	0.23
	D	0.54	0.13
1060	Drv	0.20	0.72
	A	0.20	0.22
	B	0.20	0.23
	C	0.20	0.29
1085	Dry	0.20	0.68
	A	0.20	0.20
	B	0.54	0.20
	C	0.20	0.22
	D	0.54	0.16

Material	Brass/Steel Lubrication	γ_R	μ
4140 LL	Dry	0.20	0.73
	A	0.20	0.22
	B	0.20	0.24
	C	0.20	0.24
	D	0.54	0.16
4150	Dry	0.20	0.67
	A	0.20	0.25
	B	0.54	0.19
	C	0.20	0.23
4620	Dry	0.20	0.74
	A	0.20	0.26
	B	0.20	0.27
	C	0.20	0.23
	D	0.54	0.11
5130 LL	Dry	0.20	0.62
	A	0.20	0.26
	B	0.54	0.20
	C	0.20	0.21
	D	0.54	0.10
8214	Dry	0.20	0.69
	A	0.20	0.22
	B	0.54	0.20
	C	0.20	0.21
8620	Dry	0.20	0.69
	A	0.20	0.21
	B	0.54	0.17
	C	0.20	0.20
	D	0.54	0.18
52100	Dry	0.20	0.80
	A	0.20	0.26
	B	0.54	0.20
	C	0.20	0.21
Carpenter 11 Special Steel Annealed, Water Hard	Dry	0.20	0.74
	A	0.20	0.22
	B	0.54	0.19
	C	0.20	0.19

TABLE III-3 (Cont.)

Brass/Steel			
Material	Lubrication	γ_R	μ
Hampden Steel Annealed Oil Wear	Dry	0.20	0.73
	A	0.20	0.17
	B	0.20	0.24
	C	0.20	0.21
HYCC (HA)	Dry	0.20	0.73
	A	0.20	0.21
	B	0.20	0.23
	C	0.20	0.19
	D	0.54	0.12
HYCC (PM)	Dry	0.20	0.68
	A	0.20	0.18
	B	0.54	0.21
	C	0.20	0.23
	D	0.54	0.10
Nitroalloy G	Dry	0.20	0.67
	A	0.20	0.20
	B	0.54	0.17
	C	0.20	0.22
Rexalloy AA	Dry	0.20	0.68
	A	0.20	0.15
	B	0.20	0.17
	C	0.20	0.26
Star Zenith Steel Annealed Red Wear	Dry	0.20	0.72
	A	0.20	0.22
	B	0.20	0.15
	C	0.20	0.17

Brass/Nickel Alloy			
Material	Lubrication	γ_R	μ
Carpenter Free Cut Invar "36" Annealed	Dry	0.20	0.76
	A	0.20	0.26
	B	0.54	0.20
	C	0.20	0.23
HyMu 90 Annealed	Dry	0.20	0.81
	A	0.20	0.25
	B	0.54	0.21
	C	0.20	0.23

Brass/Nickel Alloy			
Material	Lubrication	γ_R	μ
Monel C	Dry	0.20	0.85
	A	0.54	0.22
	B	0.54	0.22
	C	0.20	0.28

Brass/Copper Alloy			
Material	Lubrication	γ_R	μ
Be-Cu	Dry	0.20	0.80
	A	0.20	0.25
	B	0.20	0.23
	C	0.20	0.13
Cu-Ni	Dry	0.20	0.90
	A	0.54	0.25
	B	0.54	0.19
	C	0.54	0.24
Phos-Bronze A	Dry	0.20	0.79
	A	0.20	0.26
	B	0.54	0.21
	C	0.54	0.26

Brass/Aluminum Alloy			
Material	Lubrication	γ_R	μ
43 Aluminum	Dry	0.20	1.12
	A	0.54	0.23
	B	0.20	0.25
	C	0.20	0.20
112 Aluminum	Dry	0.20	0.92
	A	0.20	0.23
	B	0.20	0.25
	C	0.20	0.28
195 Aluminum	Dry	0.20	0.94
	A	0.20	0.17
	B	0.20	0.25
	C	0.20	0.24

TABLE III-3 (Cont.)

| Brass/Aluminum Alloy | | | | | Sintered Bronze 3/Steel | | | |
Material	Lubrication	γ_R	μ		Material	Lubrication	γ_R	μ
220 Aluminum	Dry	0.20	0.95		8620	A	0.20	0.26
	A	0.20	0.26		1117	A	0.20	0.20
	B	0.20	0.27		4140	A	0.20	0.21
	C	0.20	0.25					
355 Aluminum	Dry	0.20	1.11		Sintered Steel 1/Steel			
	A	0.20	0.22		Material	Lubrication	γ_R	μ
	B	0.20	0.24		4140	A	0.54	0.21
	C	0.20	0.25					
356 Aluminum	Dry	0.20	1.12					
	A	0.54	0.24					
	B	0.54	0.25					
	C	0.20	0.25					

TABLE III-4

DESCRIPTION OF MATERIALS

STAINLESS STEELS

302 H_m: 270 kg/mm² τ_y: 58 x 10³ psi

Composition

C	Mn	Si	Cr
.08-.20	2.00 max	1.00 max	17.0-19.0

Ni	P	S	Fe
8.00-10.00	.040 max	.030 max	Balance

303 EZ H_m: 296 kg/mm² τ_y: 63 x 10³ psi

Composition

C	Mn	Si	Cr
.15 max	2.00 max	1.00 max	17.0-19.00

Ni	P	S	Mo
8.00-10.00	.04 max	.18-.35	.60 max

Fe
Balance

321: H_m: 224 kg/mm² τ_y: 40 x 10³ psi

Composition

C	Mn	Si	Cr
.08 max	2.00 max	1.00 max	17.0-19.0

Ni	P	S	Ti
8.00-11.00	.040 max	.030 max	5 x C min

Fe
Balance

347: H_m: 252 kg/mm² τ_y: 50 x 10³ psi

Composition

C	Mn	Si	Cr
.08 max	2.00 max	1.00 max	17.0-19.0

Ni	P	S	Cb or Nb
9.00-12.00	0.040 max	0.030 max	10 x C min

Fe
Balance

STAINLESS STEELS

410: H_m: 270 ky/mm² τ_y: 58 x 10³ psi

Composition

C	Mn	Si	Cr
.15 max	1.00 max	1.00 max	11.5-13.5

P	S	Fe
.040 max	.030 max	Balance

416 EZ: H_m $\begin{cases} 270 \text{ kg/mm}^2 \\ 224 \text{ kg/mm}^2 \end{cases}$ τ_y: $\begin{cases} 58 \times 10^3 \text{ psi} \\ 40 \times 10^3 \text{ psi} \end{cases}$

Composition

C	Mn	Si	Cr
.15 max	1.25 max	1.00 max	12.00-14.00

P	S	Mv	Fe
.04 max	.18-.35	.60 max	Balance

440 C: H_m: 296 kg/mm² τ_y: 63 x 10³ psi

Composition

C	Mn	Si	Cr
.45-1.20	1.00 max	1.00 max	16.0-18.0

P	S	Mo	Fe
.04 max	.03 max	.75 max	Balance

TABLE III-4 (Cont.)

STEELS

1018: H_m: 199 kg/mm^2 τ_y: 33 x 10^3 psi

Composition

C	Mn	P	S
.15 – .20	.60 – .90	.04 max	.05 max

Fe
Balance

1045: H_m: 468 kg/mm^2 τ_y: 106 x 10^3 psi

Composition

C	Mn	P	S
.43 – .50	.60 – .90	.04 max	.05 max

Fe
Balance

1055: H_m: 270 kg/mm^2 τ_y: 58 x 10^3 psi

Composition

C	Mn	P	S
.50 – .60	.60 – .90	.04 max	.05 max

Fe
Balance

1060: H_m: 397 kg/mm^2 τ_y: 90 x 10^3 psi

Composition

C	Mn	P	S
.55 – .65	.60 – .90	.04 max	.05 max

Fe
Balance

1085: H_m: 359 kg/mm^2 τ_y: 80 x 10^3 psi

Composition

C	Mn	P	S
.80 – .93	.70 – 1.0	.04 max	.05 max

Fe
Balance

STEELS

1117: H_m: 160 kg/mm^2 τ_y: 27 x 10^3 psi

Composition

C	Mn	P	S
.14 – .20	1.00 – 1.30	.040 max	.08 – .13

Fe
Balance

4140: H_m: 180 kg/mm^2 τ_y: 32 x 10^3 psi

Composition

C	Mn	Si	Cr
.38 – .43	.75 – 1.00	.20 – .35	.80 – 1.10

P	S	Mo	Pb
.040 max	.040 max	.15 – .25	.15 – .35

Fe
Balance

4140 LL: H_m: 384 kg/mm^2 τ_y: 82.5 x 10^3 psi

Compsition

C	Mn	Si	Cr
.38 – .43	.75 – 1.00	.20 – .35	.80 – 1.10

P	S	Mo	Pb
.040 max	.040 max	.15 – .25	.15 – .35

Fe
Balance

4150: H_m: 276 kg/mm^2 τ_y: 65 x 10^3 psi

Composition

C	Mn	Si	Cr
.48 – .53	.75 – 1.00	.20 – .35	.80 – 1.10

Mo	P	S	Fe
.15 – .25	.04 max	.04 max	Balance

4620: H_m: 242 kg/mm^2 τ_y: 47 x 10^3 psi

Composition

C	Mn	Si	Mo
.17 – .22	.45 – .65	.20 – .35	.20 – .30

Ni	P	S	Fe
1.65 – 2.00	.040 max	.040 max	Balance

TABLE III-4 (Cont.)

STEELS

5130 LL: H_m: 260 kg/mm² τ_y: 55 × 10³ psi

Composition

C	Mn	Si	Cr
.27 - .33	.60 - 1.00	.20 - .35	.75 - 1.20

P	S	Pb	Fe
.04 max	.04 max	.15 - .35	Balance

8214: H_m: 220 kg/mm² τ_y: 40 × 10³ psi

Composition

C	Mn	Mo	Cr
.90	1.45	.07	.39

Si	Fe
.50	Balance

8620: H_m: 216 kg/mm² τ_y: 40 × 10³ psi

Composition

C	Mn	Si	Cr
.18 - .23	.70 - .90	.20 - .35	.40 - .60

Ni	P	S	Mo
.40 - .70	.04 max	.04 max	.15 - .25

Fe
Balance

52100: H_m: 746 kg/mm² ; 150 × 10³ psi
             220 kg/mm² τ_y: 40 × 10³ psi

Composition

C	Mn	Si	Cr
.95 - 1.10	.25 - .45	.20 - .35	1.30 - 1.60

P	S	Fe
.025	.025	Balance

Carpenter 11, Special Steel Annealed,
Water Hard: H_m: 226 kg/mm² τ_y: 40 × 10³ psi

Composition

C	Mn	Si	S
1.05	.26	.23	.009

P	Fe
.014	Balance

STEELS

Hampden Steel
Annealed, Oil Wear: H_m: 262 kg/mm² τ_y: 55 × 10³ psi

Composition

C	Mn	Si	Cr
2.17	.36	.45	11.83

Ni	P	S	Fe
.50	.019	.009	Balance

HYCC (HA): H_m: 340 kg/mm² τ_y: 75 × 10³ psi

Composition

C	Cr	Fe
1.82	11.4	Balance

HYCC (PM): H_m: 270 kg/mm² τ_y: 58 × 10³ psi

Composition

C	Cr	Fe
.10	1.68	Balance

Ketos: H_m: 296 kg/mm² τ_y: 63 × 10³ psi

Composition

C	Mn	W	Cr
.83	1.56	.07	.40

Fe
Balance

Nitralloy G: H_m: 396 kg/mm² τ_y: 90 × 10³ psi

Composition

C	Mn	Si	Cr
.42	.66	.33	1.51

Pb	P	S	Mo
.15 - .35	.015	.016	.33

Al	Fe
1.18	Balance

Rexalloy AA: H_m: 350 kg/mm² τ_y: 80 × 10³ psi

Composition

C	W	Cr	Fe
.73	.03	4.55	Balance

TABLE III-4 (Cont.)

STEELS

Star Zenith Annealed,
Red Wear: H_m: 269 kg/mm² τ_y: 58 x 10³ psi

Composition			
C	Mn	Si	Cr
.73	.26	.27	3.95
V	P	W	S
.99	.025	17.43	.011

Fe
Balance

NICKEL ALLOYS

Carpenter Free Cut
Invar "36" Annealed: H_m: 184 kg/mm² τ_y: 30 x 10³ psi

Composition			
C	Mn	Si	Se
.08	.72	.28	.15
Ni	P	S	Fe
36.07	.010	.008	Balance

Hy Mu 80 Annealed: H_m: 270 kg/mm² τ_y: 58 x 10³ psi

Composition			
C	Mu	Si	Mo
.05	.65	.15	4.0
Ni	P	S	Fe
80.0	.020	.010	Balance

Monel C: H_m: 184 kg/mm² τ_y: 40 x 10³ psi

Composition			
C	Mu	Si	Ni
.35 max	.5-1.5	1.0-2.24	62.0-68.0
Other	Cu		
2.5 max	Balance		

COPPER ALLOYS

Brass: H_m: 115 kg/mm² τ_y: 17.9 x 10³ psi

Composition	
Cu	Zn
63.5-66.5	33.5-36.5

Be-Cu H_m: 199 kg/mm² τ_y: 31 x 10³ psi

Composition			
Be	Co	Fe	Ni
1.90-2.15	.15-.50	.15-.50	.15-.50
Other	Cu		
.5 max	Balance		

Cu-Ni: H_m: 171 kg/mm² τ_y: 35 x 10³ psi

Composition			
Mn	Zn	Sn	Ni
1.0 max	1.0 max	1.50 max	29.0-33.0
Fe	Pb	Cu	
.6 max	.05 max	Balance	

Phos.-Bronze A: H_m: 166 kg/mm² τ_y: 27 x 10³ psi

Composition			
Sn	P	Pb	Fe
3.5-5.8	.03-.35	.05 max	.10 max
Zn	Sb	Cu	
.30 max	.01 max	Balance	

TABLE III-4 (Cont.)

ALUMINUM ALLOYS

43 Aluminum: H_m: 60.7 kg/mm² τ_y: 8 × 10³ psi

Composition

Si	Mg	Fe	Mn
4.5 - 6.0	.05 max	.80 max	.10 max

Zn	Cu	Ti	Al
.10 max	.10 max	.20 max	Balance

112 Aluminum: H_m: 117 kg/mm² τ_y: 15 × 10³ psi

Composition

Si	Mn	Fe	Cu
.52	.27	1.10	7.28

Zn	Ti	Al
1.24	.03	Balance

195 Aluminum: H_m: 96.8 kg/mm² τ_y: 15 × 10³ psi

Composition

Si	Mg	Fe	Cu
1.20 max	.03 max	1.0 max	4.0 - 5.0

Zn	Si	Ti	Al
.10 max	1.20 max	.20 min	Balance

220 Aluminum: H_m: 124.5 kg/mm² τ_y: 18 × 10³ psi

Composition

Si	Mg	Fe	Cu
.20 max	9.5 - 10.6	.30 max	.20 max

Zn	Mn	Al
.10 max	.10 max	Balance

355 Aluminum: H_m: 90.5 kg/mm² τ_y: 14 × 10³ psi

Composition

Si	Mg	Fe	Cu
4.5 - 5.5	.40 - .60	.60 max	1.0 - 1.5

Zn	Mn	Ti	Al
.10 max	.10 max	.20 max	Balance

ALUMINUM ALLOYS

356 Aluminum: H_m: 62.1 kg/mm² τ_y: 8 × 10³ psi

Composition

Si	Mg	Fe	Cu
6.5 - 7.5	.20 - .40	.60 max	.20 max

Zn	Mn	Ti	Al
.10 max	.1 max	.2 max	Balance

SINTERED BRASSES

Sintered Brass 1: H_m: 115 kg/mm² τ_y: 17.9 × 10³ psi

Description
7.5 min — proprietary

Sintered Brass 2: H_m: 96 kg/mm² τ_y: 14 × 10³ psi

Description
7.0 - 7.5 — proprietary

SINTERED BRONZES

Sintered Bronze 1: H_m: 135 kg/mm² τ_y: 22.5 × 10³ psi

Description
6.4 - 6.8 — ASTM B202 - 58T — Type 11 Class A

Sintered Bronze 2: H_m: 150 kg/mm² τ_y: 25 × 10³ psi

Description
7.0 min — ASTM B255 - 61T — Type 11

SINTERED IRONS

Sintered Iron 1: H_m: 180 kg/mm² τ_y: 31.5. × 10³ psi

Description
7.5 min — proprietary

Sintered Iron 2: H_m: 150 kg/mm² τ_y: 25 × 10³ psi

Description
7.3 min — proprietary

Sintered Iron 3: H_m: 110 kg/mm² τ_y: 17.9 × 10³ psi

Description
7.0 min — proprietary

TABLE III-4 (Cont.)

SINTERED IRON-COPPERS

Sintered Iron-Copper 1: H_m: 220 kg/mm^2 τ_y: 40 x 10^3 psi

Description
7.1 Copper infiltrated — 15%

Sintered Iron-Coppers: H_m: 190 kg/mm^2 τ_y: 33 x 10^3 psi

Description
5.8 - 6.2 High Copper — 20%

SINTERED STEELS

Sintered Steel 1: H_m: 220 kg/mm^2 τ_y: 40 x 10^3 psi

Description
7.0 min. Stainless AISI Type 316L

Sintered Steel 2: H_m: 150 kg/mm^2 τ_y: 25 x 10^3 psi

Description
7.0 min. — proprietary

Sintered Steel 3: H_m: 100 kg/mm^2 τ_y: 15 x 10^3 psi

Description
6.0 - 6.5 — proprietary

PLATINGS

(See Insets for hardness profiles of the platings listed)

Platings 1, 2, 3, and 4: Total Plating Thickness: 200 - 2000 μ in.

Description

These four platings differed in plating techniques. In all cases the various layers were as follows:

Substrate	First Layer	Second Layer	Other Layer
Beryllium	Copper Plating	Nickel Plating	Chromium Plating

TABLE III-4 (Cont.)

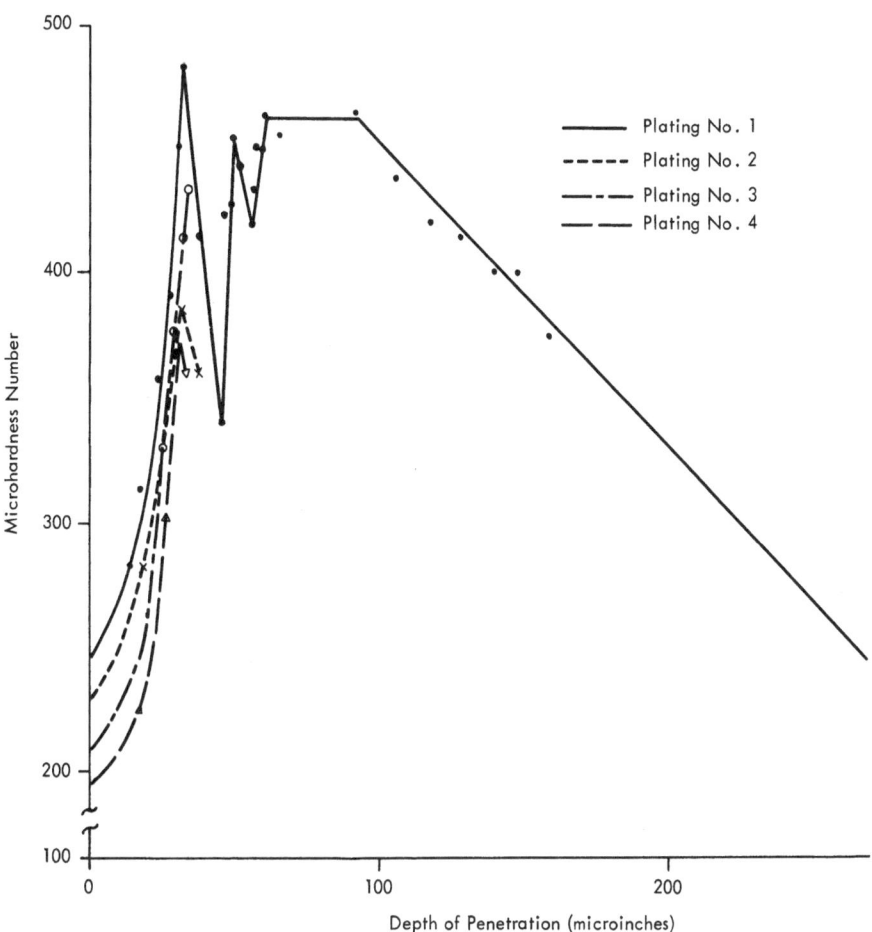

Fig. III-4 - Hardness Profile — Platings 1-4

Nickel-Cobalt Plating 1: Total Plating Thickness: 20-30 μin.   Nickel-Cobalt Plating 2: Total Plating Thickness 20-30 μin.

Description   Description

Substrate	Surface Layer		Substrate	Surface Layer
aluminum	nickel-cobalt plating		brass	heat treated nickel-cobalt plating

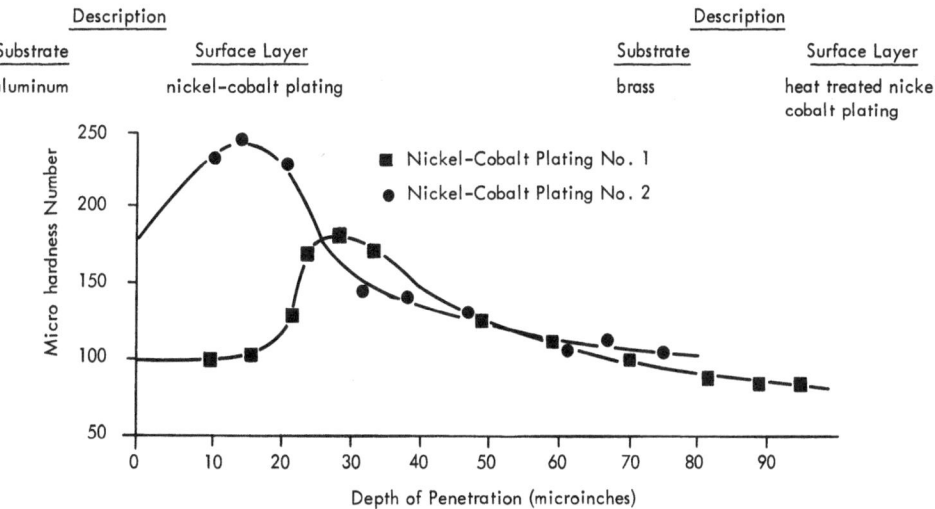

Fig. III-5 - Hardness Profile — Ni-Co Platings

TABLE III-4 (Cont.)

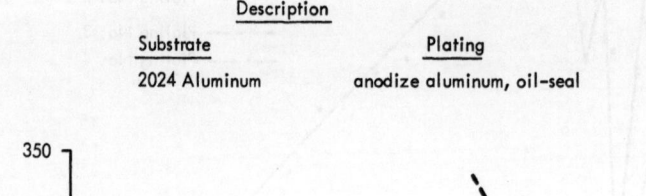

Anodized Aluminum: Total Plating Thickness, 200 – 400 μin.

<u>Description</u>

<u>Substrate</u>   <u>Plating</u>

2024 Aluminum   anodize aluminum, oil-seal

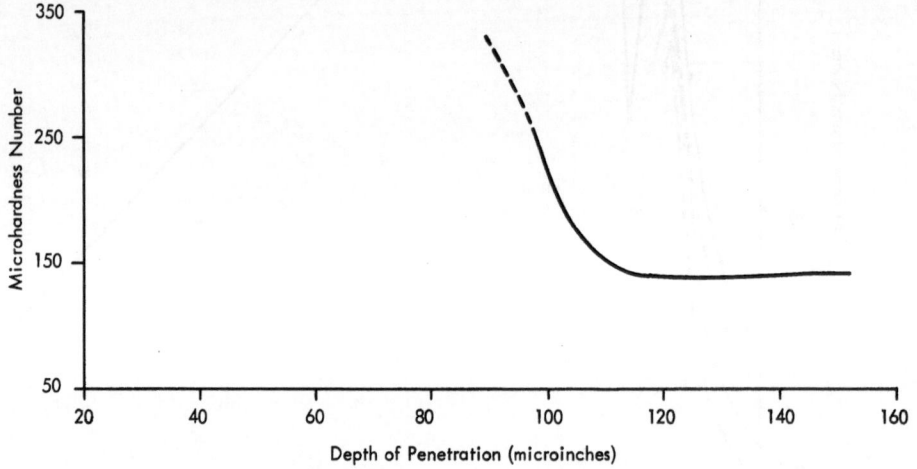

Fig. III-6 - Hardness Profile — Anodized Aluminum Platings

Chrome Plating: Total Plating Thickness, 300 – 600 μ in.

<u>Description</u>

<u>Substrate</u>   <u>Plating</u>

1018 Steel   outer layer – chromium, 10 – 100 μ in.

inner layer – nickel, 300 – 500 μ in.

TABLE III-4 (Cont.)

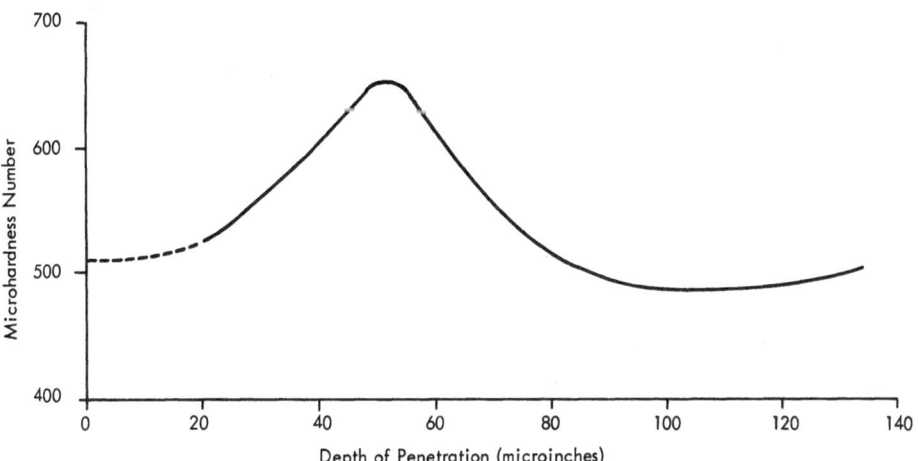

Fig. III-7 – Hardness Profile — Chrome Plating

PLASTICS

Delrin 500*: H_m: 88 kg/mm^2 τ_y: 1.235 x 10^3 psi

Description

Acetal Resin — Polymerized Formaldehyde

Nylatron G**: H_m: 88 kg/mm^2 τ_y: 1.232 x 10^3 psi

Description

Graphite Filled Nylon FM-10001

Nylatron GS**: H_m: 80 kg/mm^2 τ_y 1.14 x 10^3 psi

Description

Molybdenum Disulphide Filled Nylon FM-10001

Polyethylene 1: H_m: 20 kg/mm^2 τ_y: .172 x 10^3 psi

Description

Low Density – Sp. Gravity 0.913

PLASTICS

Polyethylene 2: H_m: 17 kg/mm^2 τ_y: .141 x 10^3 psi

Description

Medium Density – Sp. Gravity 0.926

Polyethylene 3: H_m: 8.2 kg/mm^2 τ_y: .064 x 10^3 psi

Description

High Density – Sp Gravity 0.941

Polystyrene: H_m: 45 kg/mm^2 τ_y: .535 x 10^3 psi

Description

Polymerized Styrene Plastic

Teflon*: H_m: 16 kg/mm^2 τ_y: .130 x 10^3 psi

Description

Tetrafluorolthylene Resin

Zytel 101*: H_m: 76 kg/mm^2 τ_y: 1.065 x 10^3 psi

Description

Nylon Resin

* Registered Trademark, F. T. du Pont de Nemours, Co. , Inc.,
Wilmington, Delaware.

** Registered Trademark, The Polymer Corp. , Reading, Pa.

TABLE III-5
DESCRIPTION OF LUBRICANTS

OIL A
(Socony Vaccuum Gargoyle PE 797)

Type of stock	– Paraffin
Flash Point (open cup)	– 405°F
Pour Point	– 20°F
Gravity	– 33.0 API
Viscosity Index	– 105
Neutralization No.	– 0.05
Type of Additive	– Oxidation and corrosion

OIL B
(Esso Standard Oil Millcot K-50)

Type of stock	– Napthenic
Flash Point (open cup)	– 435°F
Pour Point	– 15°F
Gravity	– 23.1 API
Viscosity Index	– 77
Neutralization No.	– 0.03
Type of Additive	– Oxidation and tackiness

OIL C
(Texaco MIL-0-5606)

Type of stock	– Paraffin
Flash Point (open cup)	– 200°F
Pour Point	– -75°F
Gravity	– 1.15 – 1.18 (specific gravity)
Viscosity Index	– 188
Neutralization No.	– 0.20
Type of Additive	– V.I. improver-anti-–wear

OIL D
Oil A Doped With 0.2% By Weight Of Stearic Acid

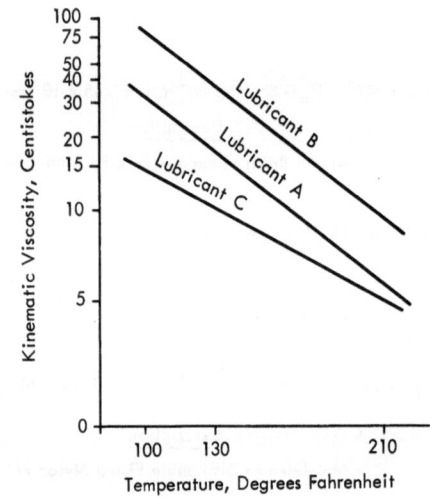

Fig. III-8 – Kinematic Viscosity For Lubricants A, B, and C.

IV REFERENCES

1. R.G. Bayer, W.C. Clinton, C.W. Nelson and R.A. Schumacher, "Engineering Model for Wear", Wear, 5, (1962), pp. 378-391.

2. R.G. Bayer, W.C. Clinton and J.L. Sirico, "A Note on the Application of the Stress Dependency of Wear in the Wear Analysis of an Electrical Contact", Wear, 7, (1964) pp. 280-287.

3. W.C. Clinton, T.C. Ku and R.A. Schumacher, "Extension of the Engineering Model for Wear to Plastics, Sintered Metals and Platings", Wear, 7, (1964), pp. 354-367.

4. A.R. Wayson, "A Study of Fretting on Steel", Wear, 7, (1964) pp. 435-450.

5. D.B. Barovich, S.C. Kingsley, and T.C.Ku, "Stresses in a Thin Strip or Slab With Different Elastic Properties From That of the Substrate Due to Elliptically Distributed Load," Int. J. Engng. Sci., 2, (1964) pp. 253-268.

6. B.W. Mott, "Micro-Indentation Hardness Testing," Butterworths Scientific Publications, London, 1956.